Mathematics Research Developments

AF848880

nova
science publishers
www.novapublishers.com

Mathematics Research Developments

Understanding Integro-Differential Equations
J. Vasundhara Devi, PhD (Editor)
Zahia Drici, PhD (Editor)
Farzana A. McRae, PhD (Editor)
2023. ISBN: 979-8-89113-040-1 (Hardcover)
2023. ISBN: 979-8-89113-136-1 (eBook)

J-Shaped Distributions and Their Applications
Mohammad Ahsanullah, PhD (Editor)
Mohammad Shakil (Editor)
2023. ISBN: 979-8-89113-013-5 (Softcover)
2023. ISBN: 979-8-89113-182-8 (eBook)

Randomness Revisited using the V Programming Language
Soubhik Chakraborty, PhD (Editor)
Subhomoy Haldar, MSc (Editor)
2023. ISBN: 979-8-89113-328-0 (Softcover)
2023. ISBN: 979-8-89113-388-4 (eBook)

Characterizations of Recently Introduced Univariate Continuous Distributions IV
G. G. Hamedani (Editor)
2023. ISBN: 979-8-88697-751-6 (eBook)

Future Relativity, Gravitation, Cosmolog
Valeriy V. Dvoeglazov, PhD (Editor)
M. de G. Caldera Cabral (Editor)
J. A. Cázares Montes (Editor)
J. L. Quintanar González (Editor)
2023. ISBN: 979-8-88697-455-3 (Hardcover)
2023. ISBN: 979-8-88697-527-7 (eBook)

More information about this series can be found at
https://novapublishers.com/product-category/series/mathematics-research-developments/

Franco F. Orsucci
and Nicoletta Sala
Editors

Complexity Science in Human Change

nova
science publishers
www.novapublishers.com

DOI: https://doi.org/10.52305/RPMQ0001

NOTICE TO THE READER

Library of Congress Cataloging-in-Publication Data

ISBN: 979-8-89113-408-9 (softcover)
ISBN: 979-8-89113-581-9 (e-book)

Published by Nova Science Publishers, Inc. † New York

Contents

Introduction

Franco Orsucci
and Nicoletta Sala*

"I think the next century will be the century of complexity."
Stephen Hawking

This book is dedicated to some of the best papers in neuroscience, psychology, and psychiatry appearing in *Chaos and Complexity Letters - International Journal of Dynamical Systems Research* (*CCL*), particularly in the last decade of the journal's life. The structure of *CCL* added value to trans-disciplinary approaches while, at the same time, differentiating the epistemology of different contributions. Empirical research, models, simulations, data analysis, and an innovative metaphors section marked the original framework of *CCL*. After twenty years, we can now say that we achieved our goals, helping to establish a common ground for a more profound and robust Complexity Science.

In this enterprise, we received support and inspiration from great companions and friends: Ilia Prigogine, Francisco Varela, Tito Arecchi, Joe Zbilut, Franco Scalzone, Walter J. Freeman, Tullio Minelli, and Mario Fulcheri. Their vision and brilliance are still alive with us. Equally, the Editorial Board has outstandingly sustained our knowledge adventures over the last twenty years in shaping this new science. Nova Science, our Publisher, has supported our journey throughout different phases. Mainly, Frank Columbus, Nadya Columbus, and Louis Maurici have been outstanding in encouraging and supporting the development and growth of the Journal.

Over the years, several special issues have been evolving into standalone books. As a result, this component of our scientific publishing has been gradually becoming more successful than the Journal itself. Therefore, we

* Corresponding Author's Email: nicolettasalavb@gmail.com.

In: Complexity Science in Human Change
Editors: Franco F. Orsucci and Nicoletta Sala
ISBN: 979-8-89113-408-9

decided to focus on developing a new Complexity Science book series, *Human Dynamics, Multidisciplinary Complexity Science*, as the Journal has completed its natural life cycle.

Book Organization

This book is divided into two sections.

The first section, *Theoretical Chapters*, contains three relevant works dedicated to complexity science in human change, some of which have already appeared in *CCL* in the past decade. Their prominence suggested to the editors to repurpose them in this book.

In the first chapter, "The Role of the Code's Conventional Rules in Complex Bio-Psycho-Cultural Systems," Massimo Agnoletti explores why *code* is necessary to understand complex systems, highlighting the importance of conventional rules. As a consequence of this perspective, any system where we can detect an increase in complexity implies the presence of specific coding structures.

Franco Scalzone and Gemma Zontini in "Roaming in Through Memory in Psychoanalysis (Mnestic Systems Dynamics)" explore the relationship between memory systems and the human experience of time as dynamic structures or processes.

In "Reconstructing Experience in Inward and Outward Organizations," Bernardo Nardi describes Personal Meaning Organizations (PMOs). It consists of the assimilation and reprocessing of experience, according to the development route of individual adaptive competencies allowing focus on the physiological resources provided by each subject's PMO.

The second section, *Technical Chapters,* collects four works that describe the connection between complexity, nonlinearity, psychology, and movement science.

In the first chapter of this section, Francesca Stecchi, and Silvia Leone, in "Chaos and Complexity of Athletes' Emotional Experiences: The Insynergy Sport Fast Emotional Insight (ISFEI) Game Cards," introduce the InSynergy Sports Emotional Insight (ISFEI Game Cards) as an explorative counseling framework. This tool can assist sports psychology practitioners in understanding how athletes experience emotions. In addition, this tool can elicit clients' deep experience narratives in non-clinical settings.

In "An Experimental Verification of NeuroPhysics Treatment by Executing a Multifractal Analysis of Surface Electromyography Signals on the

Trapezius, Abdominals, and Adductor Muscles in Athletes," Calogero Rinzivillo, Ferda Kaleagasioglu, Francesco Casciaro, Fabio Scoppa, Riccardo Marvulli, Ken Ware, Elio Conte, and Antonio Di Ieva describe a NeuroPhysics Treatment (NPT) which is an exercise-based approach that was hypothesized to utilize the body's genius to trigger and organize self-healing. NPT involves mild resistance training exercises with prolonged movements in an erect posture under therapeutic coaching. NPT was hypothetically based on system dynamics, chaos, and fractal theory. This study aimed to investigate this thesis with a cross-disciplinary approach.

In "Integrated Psychosomatic-Somatopsychic Training in a Systemic Approach for Youth Competitive Sport Technicians," Laura Cancellara and Sonia Ballocco describe a training that integrates a psychosomatic-somatopsychic approach in developing the agonistic Self. In this way, the athlete can master their skills, self-regulating in moments of emotional imbalance, and reach an optimal state of flow during performances.

In "Rorschach Ink Blots are Multifractals," Sergio Conte, Leonardo Mendolicchio, Ferda Kaleagasioglu, Rich Norman, Fang Wang, Francesco Casciaro Losito, Anna Maria Losito, Elio Conte, and Franco Orsucci examine normal subjects' physiological and perceptual reactions. The authors elaborate that the intrinsic complexity of the ink blots induces in the subjects an emotive and cognitive condition of profound inner conflict highlighted by the generalized Hurst exponent in GSR recordings. Their results could be significant in the theoretical and clinical study of cases of neurological, psychological, and psychiatric interest.

Franco Orsucci and Nicoletta Sala
Editors

Theoretical Chapters

Chapter 1

The Role of the Code's Conventional Rules in Complex Bio-Psycho-Cultural Systems*

Massimo Agnoletti†**, PhD**
University of Verona, Verona, Italy

Abstract

The human species is characterized by the particular interaction between the biological, psychological, and cultural domains which, while presenting peculiar differences, share the fact of using codes.

Each biological, psychological, or cultural domain allows for an increase in complexity based on conventional rules, characteristic of codes, which consist in establishing an arbitrary and specific relationship between an informational structure (memory space) and the result of the semantic processing of this information.

The conventional rules that characterize the codes do not emerge in a deterministic way from the properties of the substrate in which they are implemented, in fact, they are conventional rules, not physical laws.

Biological, psychological, and cultural domains, despite having the logical architecture of the codes in common, being made in different substrates give rise to different evolutionary processes.

The recent field of *Code Biology* examines the general nature of codes that can be identified within a continuum of codes represented by the genetic code at one end and the cultural code at the opposite end.

We will describe some examples of complexity promoted by codes in the biological domain (the genetic code), psychological domain (the development of language), and cultural domain (musical notes).

---

* This is a reformatted version of an article previously published in Chaos and Complexity Letters, Volume 16 Issue 3 (2022), New York: Nova Science Publishers.
† Corresponding Author's E-mail: max.agnoletti@gmail.com.

In: Complexity Science in Human Change
Editors: Franco F. Orsucci and Nicoletta Sala
ISBN: 979-8-89113-408-9

The paper will try to demonstrate why, both on a biological, psychological, and cultural level, the concept of code is a necessary requirement to explain the increase in complexity, bringing out the importance of conventional rules.

As a consequence of this perspective, any system where we can detect an increase in complexity implies the presence of a specific code in which the conventional rules are realized.

Keywords: bio-psycho-cultural model, code biology, code concept, complexity, conventional code rules

The Bio-Psycho-Cultural Model and the Code Concept

The bio-psycho-cultural (also called bio-psycho-social) model, states that the human species is characterized by the interaction of biological, psychological, and sociocultural factors so, for example, the condition of health or disease is the consequence of this complex and continuous interaction (Engels, 1977; Schwartz, 1982).

The current definition of health proposed by the World Health Organization is also consistent with the bio-psycho-cultural model (henceforth BPC), in fact, this important institution has long declared that "Health is a state of complete physical, mental and social well-being and not merely the absence of disease or infirmity" (World Health Organization, 2023).

The BPC model differs from the previous biomedical model mainly because of its highly integrated approach (which considers the wholeness of the person as a system that is not linearly reducible to the sum of its constituent parts) and the continuous and bidirectional interaction among the various biological, psychological, and cultural domains characterized by specific teleonomic systems that may converge or diverge with each other (Agnoletti, 2004).

The definition of the BPC model consists of the systematic analysis of the continuous and reciprocal interaction of biological, psychological, and cultural informational systems (Massimini et al., 1996) that change over time, following the regularities characteristic of evolutionary systems (Miller, 1970; Monod, 1970; Morin, 1985, 2005; Prigogine, 1976; Volkenstein & Chernavskii, 1978).

Human behaviors are thus ascribable to the complex interactions between the evolutionary dynamics of the biological, psychological, and cultural domains (Agnoletti, 2001; Boyd & Richerson, 1985; Campbell, 1960; Cole, 1996; Cloack, 1975; Feldman & Cavalli-Sforza, 1978; Lumsden & Wilson, 1981).

In recent decades, a considerable theoretical effort has been made by various disciplines, such as theoretical biology and physics, in trying to grasp the principles common to living systems by including those systems that, even conceptually, can demonstrate an increase in complexity that evolves over time (Barbieri, 2003; Dawkins, 1976; Miller, 1970; Monod, 1970; Morin, 1985; Prigogine, 1976).

In this paper, we will explore the concept of code, and in particular the recent field of Code Biology, as a possible common principle of the 'increase in complexity found in bio-psycho-cultural domains.

We propose the concept of code as fundamental to understanding the architecture of systems that increase their complexity.

The term "code" is commonly used to refer to the set of rules that are adopted by members of a human community, but this word possesses a broader and more general meaning.

Some examples of general definitions of code are, "A code is any system of signs or symbols that has a meaning." (Collins, 2023) or "code is a system of rules to convert information... into another form..." (WIKIPEDIA, 2023).

Codes can be defined in different ways, but, as for other scientific concepts (e.g., information), these heterogeneities must be considered as a consequence of a plurality of perspectives relating to the same common principle.

A code can be defined as a set of rules that matches two independent domains in an arbitrary and specific manner. If the arbitrary property of the codes' rules were not present, the correspondence between the two reference domains would be deterministically defined and established by the physical laws of the domains themselves.

In the context of the genetic code, for example, the arbitrariness is evident from the redundant property of the code itself, that is, from the fact that many amino acids needed to produce proteins are "translated" by a plurality of nucleic acid triplets.

The phenomenon of redundancy in the genetic code makes evident the nondeterministic relationship, dictated instead by conventional rules, between genotype and phenotype.

The need to use the term "code" to describe the relationship existing between genotype and phenotype arises precisely from the fact that between these two domains, there is no deterministic law but an arbitrary relationship defined by conventional rules.

In order to understand more clearly the fundamental difference between "rules" and "laws" (which we often use almost interchangeably in common parlance) we recall here the excellent work done by American biologist Howard Hunt Pattee, who highlights the universality, inevitability, and immutability of the laws of nature versus the arbitrariness and mutability of rules (Pattee, 1978).

While laws do not need to be embodied in a substrate or give rise to structures to make themselves explicit, rules need structures or constraints to be realized (Gazzaniga, 2018).

A code has been defined as a mapping between the objects of two independent worlds that are implemented by the objects of a third world called adaptors, or code makers, which embody conventional rules (Barbieri, 2003; Barbieri, 2006).

The Morse code, for example, expresses specific conventional rules between a group of signs composed of dots and dashes with the letters of the alphabet through a code maker represented by the human mind while the road code expresses specific conventional rules between a group of road symbols and certain driving behaviors realized by people.

In the case of a person who is making a dish by following the information in a cookbook, the person himself embodies the function of code maker because by following the instructions given in the cookbook he enacts behaviors aimed at making the dish.

The functional architecture of any code involves not only the concept of energy and information but also requires the concept of "meaning."

The information of the word "ape" for example, is measured by the bits used to differentiate the letters "a," "p," and "e," and to put them together in that order, and is the same in all languages using the same alphabetic system.

In English, the information describing the word "ape" means "monkey without a tail", while in Italian it means "insect that makes honey", in both languages it could have had any other meaning which is why it is said that the connection between the information, the sign, and its meaning has an arbitrary nature.

In themselves, words have no intrinsic meaning even though they are connoted by bits of information, they are symbolic structures to which meanings are attributed in order to define a specific correspondence between the domain of words and domain related objects and concepts.

As already mentioned, this correspondence is often expressed by saying that meanings are "arbitrary," but this is true only if we consider them individually in fact the words of a language, for example, are arbitrary when considered individually but together they globally form an integrated (conventional) system that ensures a specific connection between words and objects/concepts.

Codes and the meaning they convey are therefore also subject to constraints, but these are collective constraints due to the system of internal relationships (conventions).

In his seminal book *The Organic Codes*, Italian embryologist Marcello Barbieri describes the key features of organic codes as follows (Barbieri, 2003):

- they are rules of correspondence between two independent worlds;
- they give meanings to informational structures;
- they are collective rules which do not depend on the individual features of their structures.

The difference between information and meaning has an extremely important consequence in evolutionary terms because it implies that changes can occur in these two dimensions independently.

The Latin word "pater," for example, became "father," "père," "padre," "vater," and in this case, the order and nature of the letters changed but not the meaning attributed to the word itself. In the case of the word "ape," on the other hand, between English and Italian, change in meaning but without any change in information.

For this reason, it is necessary to make a clear distinction between "information processes," where only exchanges of energy and information take place, and "semantic processes," where appear rules that attribute meanings to structures carrying information.

The idea that life requires not only information but also meaning was proposed for the first time in 1985 by Marcello Barbieri through the publication of the book *Semantic Theory of Evolution* (Barbieri, 1985).

The recent field of Code Biology, which derives from the *Semantic Theory of Evolution* and which represents a further theoretical development, is the study of all the codes of life from the genetic code to the cultural code with the standard method of science (Barbieri, 2014).

After the discovery that many organic codes exist in living systems in addition to the genetic code (histone code, pattern code, circadian code, and many others) the theoretical concept of Semantic Biology was changed to Code Biology because it is primarily the existence of codes that reveals the existence, and the experimental proof, of meaning in the living system (Barbieri, 2015).

In summary Code Biology is based on four major theoretical frames:

- Code Biology describes the cell not as the traditional biological duality genotype-phenotype but as a trinity represented by the genotype-phenotype-ribotype, where the ribotype is the code maker of the cells. It is the ribonucleotide system that generates the rules of the genetic code.
- Code Biology claims evolution took place not only by natural selection but also by natural convention so there are two distinct mechanisms in evolution, not just one.
- Code Biology states that the complexity of the living system is determined by the number of its codes.
- Code Biology states that the fundamental process of life is codepoiesis not autopoiesis. As autopoiesis requires biological specificity and specificity comes from the genetic code, the ancestral system that came before that code could not have been autopoietic system. Those ancestral systems were engaged in the evolution of the genetic code so, according to Code Biology, the most fundamental process of life is codepoiesis not autopoiesis.

Code Biology differs from many other theories which have introduced the concept of meaning in biology, for example from biosemiotics (for the different attribution of the concept of interpretation) and from the bio-mathematician Robert Rosen's "Relational biology" (for the different attribution of the concept of anticipation) (Barbieri, 2019).

For the purposes of this writing, I propose to summarize the concept of code as a trinity consisting of:

- An informational structure (memory space) that complies with specific and defined conventional rules and serves as the informational input for the structure called code maker;

- A structure (code maker) that, by processing the input represented by the arbitrary information in the memory space, assigns meaning to them producing an output of higher complexity than the output;
- A structure that possesses higher complexity than that of the memory space (because it also implements its meaning) and that represents the product, the output, of the code maker.

From this definition of code, we can understand that, inherent in the very nature of the code, there is an increase in the complexity of the system in fact from the informational input to the output endowed with meaning we can always detect a disproportion of complexity.

Just as in the biomolecular context of the genetic code the complexity of the genotype is always less than that of the phenotype, any code is not a mere reproduction or copying between two different domains but involves the production of a more complex structure in fact, it also requires considering the semantic dimension beyond the purely informational one.

Remaining in the genetic code context, what has just been stated is evident in the difference that exists between the process of transcription (by which the information contained in DNA is enzymatically transcribed into a complementary RNA molecule), in which there is no increase in complexity, and that of translation (by which genetic information is used to produce proteins), which, being characterized by an increase in complexity, necessitates the linguistic use of the term "code."

Before proceeding by introducing the systems that increase their complexity in the biological, psychological, and cultural domains, we specify that, in general, the term complexity is used both in the qualitative and quantitative context (Edmonds, 1999; Standish, 2014).

Complexity, in the "qualitative" context, is related to our subjective ability to understand and predict the system and with the emergent proprieties derives from the interactions from parts that make up the system (Conway, 1982; Fromm, 2004; Holland, 1997; Standish, 2014).

In the "quantitative" context, complexity is related to the objective number of information that the system required to be described (Bonner, 1988; McShea, 1996; Marcot & McShea, 2007).

In general, we could define complexity as the property of a system which makes it difficult to predict its global behavior, in a given system of representation, even with reasonably complete knowledge of its atomic components and their interrelationships (Edmonds, 1999).

Now that we have defined the concept of code, particularly in the context of Code Biology, let us see how it can be applied in the various biological, psychological, and cultural domains in which we can detect increased complexity.

The Complexity and the Conventional Rules in the Biological Domain

Paradigmatic examples of increased complexity in the biological context are embryonic development and the genetic code.

In order to understand the example of the genetic code in depth, it is useful to introduce the concept of the self-replicating machine, the minimum functional architecture that a cell must possess in order to replicate itself and thus be considered "living."

The celebrated Hungarian mathematician John von Neumann was a pioneer in the study of the logical architecture inherent in any entity that satisfies the condition of replicating itself in fact, we currently call "von Neumann architecture" his valuable contribution, which, as we shall see, also enabled a deeper understanding of how living systems work.

Extraordinarily anticipating the revolutionary discovery of the mechanism of the genetic code of biological cells, von Neumann predicted that a self-replicating system must necessarily include an informational memory with a linear order, one-dimensional, structure (von Neumann, 1966).

In addition to the specific nature of this memory, the architecture theorized by von Neumann also postulated that memory should possess two functionally different categories of information.

The first category consists of the information describing the system that is to be replicated while the second category concerns the information needed to describe the structure, the so-called Universal Turing Constructor, which, by interpreting the information of the first category, assigns meaning to them thereby concretely realizing the replication described by the same information.

As elegantly described by the South African Nobel Prize-winning biologist Sydney Brenner, von Neumann understood the logical necessity of including within the main memory also the description of a structure of the Universal Turing Constructor essential for "reading" the information related to the first category, providing them with meaning (Brenner, 2012).

This is not the same as stating that main memory includes the meaning of information itself, as erroneously claimed by the Austrian physicist Erwin Schrödinger (Brenner, 1994).

In other words, von Neumann was the first to identify very clearly the need to distinguish, at the level of memory space, between information about the program and information about the description of the system that has to "read" them, giving them meaning.

The characteristics identified by von Neumann are profoundly and remarkably overlapping both with the properties of the genetic code identified later with the discovery of DNA and with the assumptions defined by Barbieri's Code Biology, in fact, two scholars of artificial life, Daniel Mange, and Moshe Sipper describe this evident theoretical convergence by stating that:

Finally, Barbieri proposed a new theory of evolution, based on the trinity genotype-ribotype-phenotype (Figure 1). Ribosomes are capable, in general, of translating RNA chains of any length into proteins; in particular, they are capable of decoding a specific RNA string, the ribosomic RNA, producing an exact copy of the ribosome itself, an archetypical self-replication process (Figure 2).

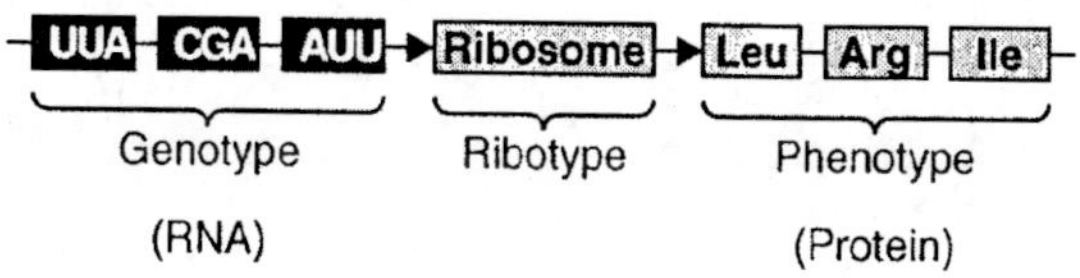

Figure 1. Self-replication of the ribosome.

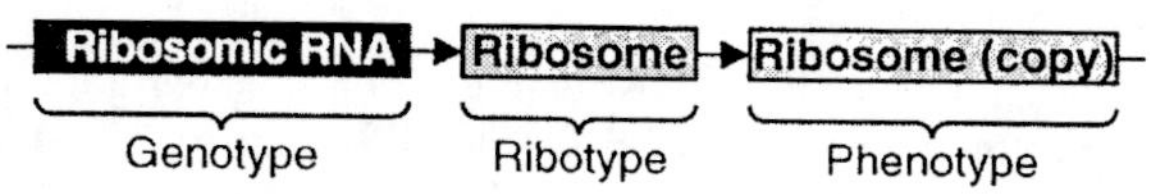

Figure 2. Self-replication of the ribosome.

Von Neumann's visionary work, carried out in the late 1940s, predates that of Crick, Roberts, Barbieri, and other biologists. We hold that his basic message with respect to self-replicating automata concerns the underlying architecture of the universal constructor—which is none other than the artificial version of the biological ribosome. One can discern the genotype-ribotype-phenotype trinity in von Neumann's cellular-automaton world (Figure 3a):

- The genotype is the (input) tape of the automaton, containing the description (genome) of the machine to be built.
- The ribotype is the universal constructor itself.
- The phenotype is the ultimate construction, in the cellular space, of the machine described on the tape.

Self-replication of the universal constructor occurs in analogy to nature (Figure 3b): The description (genotype) written on the input tape is translated via a ribosome (ribotype) so as to create the offspring universal constructor (phenotype).

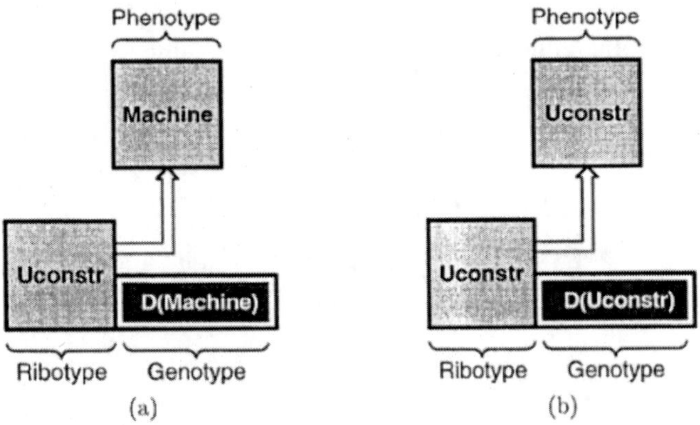

Figure 3. (a) Von Neumann's universal constructor within the genotype-ribotype-phenotype framework. (The universal constructor is a machine capable of constructing, through the use of a "constructing arm," any configuration whose description can be stored on its input tape.) (b) Self-replication of the universal constructor (Mange & Sipper, 1998, p. 226-227).

To conclude, we maintain that von Neumann's quintessential message is Genotype, Ribotype, Phenotype."

In other words, the minimum architecture required for a replicating entity such as a cell needs the existence of a code that allows biological specificity between the descriptive information contained in the memory space that is also heritable (the genotype) and its physical realization (the phenotype), through a structure (the ribotype) that using conventional rules provides meaning to the information in the memory space.

The increase in complexity expressed by the genetic code is thus to specifically relate the one-dimensional, digital, chemically inert memory space of the genotype to the three-dimensional, chemically active domain of the phenotype.

Any protein has quantitatively and qualitatively greater complexity than its informational description represented in DNA sequence, so "translation" from the one-dimensional information domain of the genotype to the three-dimensional information domain of the phenotype can be considered an epigenetic increase in complexity expressed by the genetic code (Barbieri, 2003).

Another way of interpreting this concept is that the production of a protein amounts to a process of reconstruction from incomplete information contained in the genotype in the sense that the one-dimensional memory space represented by DNA is not mathematically sufficient to describe the structural complexity of proteins.

It is only through the function of "reading" the incomplete information in the genotype and assigning a specific meaning by the ribotype that the production of the phenotype takes place.

This is why the use of the term "genetic code" referring to the context of "translation" from genetic information into that of proteins is not just a metaphor but a real entity if not the only way to describe this phenomenon (Barbieri, 2003; Barbieri, 2006).

As we have seen above, codes are characterized by the realization of conventional therefore arbitrary rules promoted by code makers.

In the biological context, the code maker is represented by the ribosome that realizes the conventional rules of the genetic code by putting genotype information into a specific relationship with phenotype information.

From the point of view of evolutionary dynamics, it is interesting to note that in this specific context, the code is realized in an implementation in which:

- the informational memory, the code maker, and the code maker product are spatially very close and
- the access to the informational memory is constrained to one code maker at a time.

In the biological context of genetic coding, the intergenerational transmission of information from one replicator to another is reserved for memory space (genotype), neither code maker (ribotype) nor code maker product (phenotype) is transmitted intergenerationally.

If the genotype is the locus of intergenerational transmission of hereditary information, the phenotype (and not the genotype) is the three-dimensional locus of physical interaction, of molecular behavior that will be subject to natural selection and thus determine the replicator's overall fitness.

Of course, in addition to genetic code, there are many other biological examples where we can observe an increase in complexity.

Another of the most striking examples of increased biological complexity is embryonic development, where the progressive exponential increase from a single cell to the production of billions of highly differentiated cells testifies to the massive and coordinated interaction of many different types of biological codes.

Let us explore an example of a code maker in which we can detect increased complexity in the human psychological domain.

The Complexity and the Conventional Rules in the Psychological Domain

In the psychological domain, certain phenomena such as creativity, the mechanism of Flow experiences, and human language development represent paradigmatic examples of increased complexity.

Language development is a process in which it is possible to detect an increase in complexity for several reasons, but the main one consists of the evident disproportionality between the complexity of the linguistic input required to learn a language versus the superior complexity of the output represented by the phenomenon of the grammar development (which is a different phenomenon from language production).

Particularly consistent with the concept of code proposed in this article is the theoretical proposal of language development by Noam Chomsky (Chomsky, 1980) concerning universal, or generative, grammar, that is, the set of rules capable of describing and explaining the learning of the grammatical structure of every possible natural language.

Generative grammar is an elaboration process that, starting from relatively few informational elements, can generate a language in its infinite grammatical sequences and establish the grammaticality of perceived and produced statements.

The basic assumption of this *innatist* perspective is that there is a uniformity, an invariance, of linguistic competence in human beings independent of the linguistic environment to which they are exposed.

Generative grammar predicts that there is a specific neural architecture, the Language Acquisition Device (LAP), that enables children to acquire language input through experience as well as produce language output.

This hypothesis states that language is a set of specific parameter values in an invariant system of principles (Chomsky, 1995) realized by the LAP.

In favor of this hypothesis is the *poverty of stimulus argument*, which argues that unless children have significant innate knowledge of grammar, they would not be able to learn the language, since the learning process is partly independent of the exposure to language input.

The richness of the language environment to which a child is exposed is thus partly independent of the ability to learn the language itself.

The *poverty of the stimulus* states that the linguistic input that children process as they experience the language exposition does not contain enough information to develop a deep understanding of the grammar of the language itself.

The different informational format between the language input order and the format characterizing Chomsky's generative grammar, linear and hierarchical respectively, emphasizes the specific role of processing and increasing the complexity of the LAD to produce a syntactic generation that is not reducible simply to the linguistic input data.

This discrepancy between the linear order of the language input and the hierarchical structure of the generative grammar can be expressed from the state that "in no language can a rule be based on the number of words in a sentence or on a specific position in the linear sequence of words" (Moro, 2018, p. 276).

The *poverty of the stimulus* argument is seen as evidence against the empiricist perspective that language is learned exclusively through a general deductive process, applied to a mental "tabula rasa," aimed at extrapolating invariances arising from experience from boundless linguistic arbitrariness.

The empiricist perspective doesn't require an increased complexity language system as the amount of linguistic input is assumed to be sufficient to explain the acquisition of the language itself.

In other words, the *empiricist* perspective assumes that by processing a sufficient amount of linguistic data, a general neural learning process can extrapolate the grammatical invariance of any language.

In contrast, the theoretical proposal of generative grammar involves that the linguistic experience required to provide information to the LAD architecture is fundamental, though not sufficient, for the language learning process thus the output represented by the generative grammar is not deducible from the linguistic input.

In addition to the principle of necessity and no sufficiency language experienced, the context of generative grammar requires that linguistic input be provided in a linear order and with an arbitrary dimension (which is not to state that the arbitrary nature of the input is absolute). That is, the format of linguistic information representing the experiential input must correspond to recognition rules through which the neural architecture of the LAD distinguishes between possible natural languages and impossible languages to process (Embick, et al., 2000; Moro et al., 2001; Musso, et al., 2003).

Only within the boundaries of the LAD's recognized natural language rules, we can find an almost infinite arbitrary dimension.

This perspective is very far from the empiricist approach that accepts that "languages [can] differ from each other without limit and in unpredictable ways" (Joos, 1957, p. 96).

At this point, it is appropriate to return to the concept of code adopted in this paper because there is a strong similarity between the theoretical proposal of generative grammar and that of codes (Barberi, 2010). In both cases, the functional architecture involves a trinity consisting of an informational input in sequential order, a structure that processes the input, and an output that represents the result of processing the initial information by attributing to it a specific meaning determined by the nature of the processor itself.

In both cases, there is an arbitrary and conventional dimension visible both in the informational format of the input and in the rules adopted by the processor that establishes the relationship between input and output.

From the evolutionary point of view, the human language learning code is realized in an implementation in which the informational memory input is:

- spatially separated from the code maker and
- the access to the memory space is not constrained to one code maker at a time (the language experience can represent an input for two or more code maker at the time).

In addition to these features, the context of the human linguistic learning code allows the linguistic communicative output (which requires a further transformation in linear order than the hierarchical structure of the generative grammar) to represent an informational input both for the same person and for other people.

Increased complexity in the context of human language predicts the need for conventional rules, established biologically through human neural architecture, that constrain the space of cultural arbitrariness of language.

In other words, the dimension of arbitrariness is present both in the conventional rules of generative grammar and in the cultural choices allowed within the space constrained by the biological code of generative grammar.

It turns out to be interesting to note that the paradigm of codes applied to Chomsky's generative grammar represents a valuable option to be explored in order to overcome the well-known discussion concerning the possibility that natural selection could have biologically generated LAD.

The trinity of elements characterizing the code makes it possible to overcome the age-old sterile nature/environment, innate/acquired dichotomy by providing not only the mechanism of natural selection but also that of natural conventions.

Before the generation of human language, there were two types of entities, living organisms and nonliving natural objects such as, for example, stones, clouds, and seas.

After the definition of the human language code, a third category of entities also appeared, cultural artifacts such as, for example, knives, books, and bikes that have a behavior that cannot be completely overlapped with either living or nonliving entities.

Studying the behavior of some artifacts such as, for example, the car, some scholars found that the trend of their diffusion within a given ecological niche followed the Lotka-Volterra equation typical of the dynamics of living things (Marchetti, 1980).

The human linguistic code allowed the generation of a new hybrid form of entities, the artifacts, which triggered the human cultural phenomenon.

Let us now look the last of the domains predicted by the bio-psycho-cultural model where we can see an increase in complexity: the cultural domain.

The Complexity and the Conventional Rules in the Cultural Domain

This paper does not pretend to be exhaustive of the numerous dynamics of culture but only to point out the potential usefulness of using the paradigm of codes in interpreting certain cultural aspects.

It is currently generally accepted that at least two types of codes exist in nature: the genetic code and the cultural code, but this does not mean either that they are the only existing codes or that there is no causal and evolutionary relationship between them (Barbieri, 2018).

On the contrary, taking it for granted that only these two codes exist is equivalent to assuming that exclusively at two historical moments separated temporally by almost four billion years, that is, at the opposite poles of all biological evolutionary history, the conditions that led to their generation were realized.

In other words, this perspective considers the genetic and cultural codes as "frozen accidents" that have no causal and evolutionary relationship with each other.

The idea that there is an insurmountable barrier between biology and culture seems to be solid and irrefutable; the extremely fast, fickle, and highly arbitrary change characteristic of cultural processes seem to be particularly distant from the logic typical of biological processes.

Biology possesses a code that has remained unchanged for millions of years, while culture has countless codes that are constantly changing and seem to be dominated by extreme arbitrariness.

If one assumes the thesis proposed by this paper, based on the concept of code referring to Code Biology, it will become evident that the perspective that both genetic and cultural codes are "frozen accidents" turns out to be unrealistic as well as illogical from an epistemological point of view.

For more than half a century, many theorists have studied the invariances of the cultural evolutionary system (Campbell, 1960; Cloak, 1975; Dawkins, 1976; Lumsden, Wilson, 1981; Massimini et al., 1996; Mesoudi et al., 2006).

One definition of cultural evolution that can be considered for the purposes of this text is that of cultural evolution as a historical, transgenerational process of transmission over time of information deposited in extra-somatic site (Massimini et al., 1993; Csikszentmihalyi, 1993).

The expression "information deposited in extra-somatic site" should be understood here in both an "absolute" and "relative" sense.

That is, in the "absolute" case it is meant that cultural information is found deposited externally with respect to the nervous system of any human subject.

Consider, for example, the construction project of an airplane, an automobile or a skyscraper; no human being individually holds this information entirely but can exist a physical memory where all this information can be collected and made accessible.

Instead, we speak of "relative", referring to the expression in object, to emphasize the fact that the "extra-somatic" of one subject may coincide with the "intra-somatic" of another subject.

To clarify this point, consider for example the oral tradition of certain poems.

For a poet, one's poem would acquire the cultural connotation only if it were declaimed to someone else (or, of course, if it were memorized in some shared artifact, as provided by the "absolute" meaning).

In general terms, it is assumed that there is a cultural evolutionary process where the information processed by the human mind accumulates externally than the human organism itself.

The process of cultural information accumulation is detectable in extra-somatic physical sites because the lifespan of intra-somatic cultural information is limited by the longevity of the person who possesses it.

For this reason, the adjective "transgenerational" is often referred to in the definition of cultural evolution.

As is evident, the condition of being deposited externally to the human neural memory system implies that there is a code by which cultural information is expressed and is also potentially accessible by other individuals.

There is currently no unanimous agreement that potentially accessible information is an integral part of the cultural system (Boyd & Richerson, 1985; Fog, 1999).

While there are historical instances whereby code became from exclusively individual to collective knowledge acquiring its social and cultural connotation, we can imagine the possibility of a code remaining secret, inaccessible, forever.

In the psychological sciences, the discussion concerns the possibility of the individual code becoming a shared, hence social, code.

If the transition from individual to shared code does not occur, cultural information may no longer be subject to evolution because it lacks the mechanism of interpretation (Csikszentmihalyi, 1988, 1996; Gardner, 1993).

If, for example, the translation engraved on the Rosetta Stone had never been decoded, we could hardly speak of cultural evolution.

From this brief digression, the importance of the social perspective in the analysis of cultural phenomena is once again evident.

In particular, the difference between individual codes and social codes appears crucial.

In this regard I find illuminating the definition of Romanian social psychologist Serge Moscovici relatively social representations understanding them as "cognitive systems with a logic and rationale of their own. ... *[which also have the function of] ... making communication between members of a community possible by providing them with a code to discriminate and classify unambiguously the various aspects of their world and their individual and group history"* (Moscovici, 1969, p. 12).

Returning to the definition of cultural evolutionary process, the basic units of this system are artifacts understood as any form of artificial human product whether material or immaterial such as objects, institutions, values, norms, laws, etc. (Massimini et al., 1996).

Indeed, artifacts fulfill the criterion of transgenerational replication of information that identifies an evolutionary system.

The teleonomy characteristic of cultural evolutionary systems requires a transmission of information that can only occur through the application of forces from biological systems, that is, from the selective investment of human attention (Csikszentmihalyi, 1978; Kahneman, 1973).

Unlike living systems, artifacts deprived of this external resource would inevitably suffer entropic degradation characteristic of nonbiological objects.

The absence of the criterion defined by French Nobel Prize-winning biologist Jacques Monod as "autonomous morphogenesis" makes clear the difference between the biological and cultural levels, in the sense that the latter deficits of "almost total freedom toward external agents and conditions" (Monod, 1970, p. 22) characteristic of biological replicators.

Any artifact is preserved and transmitted over time only if individuals turn their attention to it, selecting it, maintaining it and spreading it within the cultural niche.

A specific portion of human culture is preserved and transmitted over time only if it is able to remedy the criterion of autonomous morphogenesis by attracting the selective investment of human attention (Csikszentmihalyi & Rochberg-Halton, 1981; Massimini et al.,1996).

Italian embryologist Marcello Barbieri writes in this regard:

> ... artifacts originate in the human mind as mental objects, and afterwards are turned by man into physical objects (this is true even for a poem which must become ink on paper or soundwaves in the air). Artifacts have therefore a genotype and a phenotype. [...] Humans artifacts, in short, have a genotype and a phenotype, this qualifies them as "organisms"

> but they are organisms of a very peculiar breed, because their phenotypes are totally dissociated from their genotypes.
>
> They truly are a new form of life, and their appearance on Earth was a real episode of macroevolution. (Barbieri, 2003, p. 228)

The result of different kind of implementations between cultural and biological codes is that cultural evolution is extremely faster than biological evolution.

The emergence of culture, in short, was a true macroevolutionary phenomenon, as produced a new form of life, a new type of replicators where genotypes are physically separated from phenotypes.

The code paradigm proposed by Code Biology thus asserts that biological evolution and cultural evolution are both based on the evolution of codes, and in this sense the two phenomena are homologous, but this does not mean that between biology and culture there are no differences between the two processes.

Adopting the definition of code proposed in this paper, human cultural evolution takes its start from a code maker represented by the human mind which has established specific conventional relationships between the memory space of artifacts and their physical realization.

From an evolutionary perspective, the human cultural code is realized in an implementation in which:

- the memory space can be either close to or spatially distant from the code maker;
- access to the memory space is not limited to one code maker at a time (the design of an artifact can represent informational input for two or more code makers at a time);
- the code maker's output can be either close to or spatially distant from the code maker itself;
- the output of one code maker can represent an input memory space for another (or the same) code maker.

It can be inferred from this reflection that the evolutionary dynamics typical of cultural codes have different spatial and temporal properties than biological or psychological ones.

Just to give some examples of complexity in the domain of cultural artifacts we can think of the use of money or the making of the musical pentagram.

Regarding the graphic system of sound annotation of notes (pentagram), it is interesting to note that until the 9th century, the transmission of music took place almost only orally because earlier graphic systems that relied on solely alphabetical information were too complex and inaccurate in representing musical sounds.

From the ninth century, the first attempts at graphic representation began to appear, which, in addition to alphabetical information, included graphic signs, of notes, to capture the musical melody.

This representational code became more and more informative and precise until, around 1025, the Italian Benedictine monk Guido "D'Arezzo" introduced into the code the system of four lines (evolved to the current five, "penta," "five," -gram, that is, consisting of five lines) and the conventional rule matching sounds to the syllables of a chant well known at the time.

Guido "D'Arezzo" made a much simpler and more informative musical graphic system than the previous ones for learning, memorizing, and transmitting music, enabling more effective learning and dissemination of music.

The Italian monk promoted a true cultural revolution because, in addition to making cultured music accessible to a breadth of people, he made it easier for all generations to follow to process, produce, and disseminate music.

Guido "D'Arezzo" can be regarded as the code maker of the currently used graphical musical representation who defined the conventional rules of this code from the information of previous musical codes, increasing complexity through their elaboration.

In this brief example, the dynamics related to the fact that the informational plane that serves as input for one code maker can be considered as the output produced by another code maker so the fundamental aspect of social interactions of cultural dynamics is evident.

The Role of Conventional Code Rules in Bio-Psycho-Cultural Evolutionary Complexity

Throughout this paper, we hope that the role of conventional rules underlying the evolutionary processes that characterize the human biological, psychological, and cultural domains will be evident.

The concept of code with its evolutionary implications in no way denies the role of the evolutionary mechanism of natural selection, but it underscores the importance, and epistemological necessity, of also considering the evolutionary mechanism that determined the establishment of biological, psychological, and cultural code's conventions.

Any evolutionary dynamic, from biological to psychological and cultural, necessarily requires the existence of a replicative entity that is the object of the selective processes that determine differential fitness within any population of replicators.

We have seen that any replicative entity necessarily requires the existence of a code that conventionally and specifically establishes the relationship between an informational space and one where, to this information, is added a meaning attributed by a code maker.

In a theoretical scenario where Code Biology no longer considers genetic and cultural codes as "frozen accidents" but real protagonists of human and natural evolutionary history (Barbieri, 2003), we can state that:

- Codes consist of a trinity of factors in which there is an information space that serves as input to a code maker who, by attributing conventional meaning to this information produces as output a product of higher complexity than the input;
- codes are the fundamental invariants of any evolutionary process;
- Globally, codes possess the cumulative property in the sense that the generation of new codes never abolished previous codes by adding an ever-increasing increase in complexity.
- The rate of appearance of organic codes has been increasing throughout the history of life, and the most complex life forms are also those with the greatest diversification.

As a consequence of these statements, any system where we can detect an increase in complexity implies the presence of a specific code in which the conventional rules are realized.

References

Agnoletti, M. (2004). *Il modello bio-psico-culturale* [*The bio-psycho-cultural model*]. Dipav,11, 6-12. Franco Angeli.

Agnoletti, M. (2001). *Analisi comparativa di alcuni paradigmi utilizzati in psicologia evolutiva, biologia e vita artificiale* [*Comparative analysis of some paradigms used in evolutionary psychology, biology and artificial life*]. Master's thesis dissertation, Padua University.

Barbieri, M. (2019). Code Biology, Peircean Biosemiotics, and Rosen's Relational Biology. *Biol Theory*, 14, 21–29. https://doi.org/10.1007/s13752-018-0312-z.

Barbieri M. (2018). What is code biology? *Bio Systems*, *164*, 1–10. https://doi.org/10.1016/j.biosystems.2017.10.005.

Barbieri, M. (2015). *Code Biology. A New Science of Life*. Springer, Dordrecht.

Barbieri, M. (2014). From biosemiotics to code biology. *Biol. Theory*, 9, 2, 239–249.

Barbieri, M. (2010). On the Origin of Language. *Biosemiotics, 3*, 201-223.

Barbieri, M. (2006). Life and Semiosis: the real nature of information and meaning. *Semiotica*, 158, 233–254.

Barbieri, M. (2003). *The Organic Codes. An Introduction to Semantic Biology*. Cambridge University Press, Cambridge, UK.

Barbieri, M. (1985). *The Semantic Theory of Evolution*. Routledge.

Bonner, J. T. (1988). *The evolution of complexity*. Princeton UP.

Boyd, R., & Richerson, P. J. (1985). *Culture and the Evolutionary Process*. University of Chicago Press.

Brenner, S. (2012). Life's code script. *Nature*, 482, 461.

Brenner, S. (1994). Schrödinger wrong, von Neumann right. Retrived from: https://www.youtube.com/watch?v=a4_JfLpZpnI&list=PLeYoYWSU6ZYxMwEUKkrwyqQAJ8aTOveVR&index=2.

Campbell, D. T. (1960). Blind Variation and Selective Retention in Creative Thought as in Other Thought Processes. *Psychological Review*, 67, 380-400.

Chomsky, N. (1995). *The Minimalist Program*. MIT Press, Cambridge MA.

Chomsky, N. (1980). *Rules and representations*. New York: Columbia University.

Cloak, F. T. (1975). Is a cultural ethology possible? *Human Ecology*, 3, 161-182.

Cole, M. (1996). *Cultural psychology*. Cambridge: Harvard University Press.

Collins (2023). *Code*. In… https://www.collinsdictionary.com/it/dizionario/inglese/code.

Conway, J. H. (1982). What is life? In E. Berlekamp, J. H. Conway, & R. Guy (Eds.), *Winning ways for your mathematical plays* (Vol. 2). Academic.

Csikszentmihalyi, M. (1996). *Creativity. Flow and the Psychology of Discovery and Invention*. New York: Harper-Collins.

Csikszentmihalyi, M. (1993). *The Evolving Self. A Psychology for the Third Millennium*. New York: Harper-Collins.

Csikszentmihalyi, M. (1988). Society, Culture, Person: A system View of Creativity. In R. J. Sternberg (Cur.). *The Nature of Creativity*. New York: Cambridge University Press.

Csikszentmihalyi, M. (1978). Attention and the Holistic Approach to Behavior. In K. S. Pope, J. L. Singer (Editors), *The Stream of Consciousness*. New York: Plenum.

Csikszentmihalyi, M., Rochberg-Halton, E. (1981). *The Meaning of Things*. Cambridge University Press.

Dawkins, R. (1976). *The Selfish Gene*. Oxford University Press.

Edmonds, B. (1999). *Syntactic measures of complexity*. PhD thesis, University of Manchester. Retrieved from http://bruce.edmonds.name/thesis/.

Embick, D., Marantz, A., Miyashita, Y., O'Neil, W., & Sakai, K. L. (2000). A syntactic specialization for Broca's area. *Proceedings of the National Academy of Sciences of the United States of America, 97*(11), 6150–6154. https://doi.org/10.1073/pnas.100098897.

Engel, G. L. (1977). The need for a new medical model. A challenge for biomedicine. *Science.* 196, 129-136.

Feldman, M. W., & Cavalli-Sforza, L. L. (1984). Cultural and biological evolutionary processes: gene-culture disequilibrium. *Proceedings of the National Academy of Sciences, 81*(5), 1604-1607.

Fog, A. (1999). *Cultural Selection*. Dordrecht: Kluwer.

Fromm, J. (2004). *The emergence of complexity.* Kassel UP.

Gardner, H. (1993). *Creating Minds. An Anatomy of Creativity Seen Through the Lives of Freud, Einstein, Picasso, Stravinsky, Eliot, Graham and Ghandi*. New York: Basic Books.

Gazzaniga, M. (2018). *The Consciousness Instinct.* Farrar, Straus and Giroux.

Holland, J. (1997). *Emergence: From chaos to order.* Addison Wesley.

Kahneman, D. (1973). *Attention and Effort*. Englewood Cliffs: Prentice-Hall.

Joos, M. (Ed.) (1957). *Readings in linguistics: The development of descriptive linguistics in America since 1925*. Washington, DC: American Council of Learned Societies.

Lumsden, C., & Wilson, E. O. (1981). *Genes, Mind, and Culture*. Harvard University Press.

Mange, D. e Sipper, M. (1998). Von Neumann Quintessential Message: Genotype + Rybotype = Phenotype. *Artificial Life*, 4(3). 225-227.

Marchetti, C. (1980). Society as a learning system: Discovery, invention, and innovation cycles revisited. *Technological Forecasting and Social Change,* 18, 4, 267-282.

Marcot, J. D., & McShea, D. W. (2007). Increasing heirarchical complexity throughout the history of life: Phylogenetic tests of trend mechanisms. *Paleobiology*, 33, 182–200. doi:10.1666/06028.1.

Massimini, F., Inghilleri, P., & Delle Fave, A. (1996). *La selezione psicologica umana. Teoria e metodo d'analisi* [*Human psychological selection. Theory and method of analysis*]. Milano: Cooperativa Libraria IULM.

Massimini, F., Toscano, M., & Inghilleri, P. (1993). La selezione culturale umana [Human cultural selection]. In F. Massimini & P. Inghilleri (Editors), *La selezione psicologica umana. Teoria e metodo di analisi* [*Human psychological selection. Theory and method of analysis*]. Milano: Franco Angeli.

Mesoudi, A., Whiten, A., & Laland, K. N. (2006). Towards a unified science of cultural evolution. *The Behavioral and brain sciences*, 29(4), 329–383. https://doi.org/10.1017/S0140525X06009083.

McShea, D. W. (1996). Metazoan complexity and evolution: Is there a trend? *Evolution; International Journal of Organic Evolution*, 50, 477–492. doi:10.2307/2410824.

Miller, J. G. (1970). *Living Systems.* New York: Mc-Graw-Hill.

Monod, J. (1970). *Le hazard et la necessité*. Parigi: Seuil.

Morin E. (2005). *Introduction à la pensée complexe*. [*Introduction to Complex Thinking]* Paris, Seuil.

Morin, E. (1985). La via della complessità [The way of complexity]. In G. Bocchi, M. Ceruti (Editors) (1985), *La sfida della complessità* [*The complexity challenge*]. Milano: Feltrinelli.

Moro, A., Tettamanti, M., Perani, D., Donati, C., Cappa, S. F., & Fazio, F. (2001). Syntax and the brain: disentangling grammar by selective anomalies. *NeuroImage*, 13(1), 110–118. https://doi.org/10.1006/nimg.2000.0668.

Moro, A. (2018). *The Boundaries of Babel*. MIT Press.

Moscovici, S. (1969). Preface. In C. Herzlich (1969), *Santè e maladie*. Paris: Mouton.

Musso, M., Moro, A., Glauche, V., Rijntjes, M., Reichenbach, J., Büchel, C., & Weiller, C. (2003). Broca's area and the language instinct. *Nature neuroscience*, 6(7), 774–781. https://doi.org/10.1038/nn1077.

Pattee, H. H. (1978). The complementarity principle in biological and social structures. *Journal of Social and Biological Structures*,1, 2, 191-200. https://doi.org/10.1016/S0140-1750 (78)80007-4.

Prigogine, I. (1976). Order Through Fluctations. Self-Organization and Social Systems. In Jantsch, E. & Waddington L. H. (editors), *Evolution and Consciousness. Human Systems in Transition*. Reading (Mass.). Addison-Wesley.

Schwartz, G. E. (1982). Testing the biopsychosocial model: The ultimate challenge facing behavioral medicine? *Journal of Consulting and Clinical Psychology*, 50(6): 1040-1053.

Standish, R. K. (2014). Complexity: Quantity or Quality. *International Journal of Signs and Semiotic Systems* (IJSSS), 3(1), 27-45. http://doi.org/10.4018/ijsss.2014010103.

Volkenstein, M. C., & Chernavskii, D. S. (1978). Information and Biology. *Journal of Social and Biological Structures*, 1,1, 69-86.

Von Neumann, J. (1966). *Theory of self-reproducing automata*. Urbana: University of Illinois Press. Edited and completed by A. W. Burks.

Wikipedia (2023). Code. In… https://en.wikipedia.org/wiki/Code.

World Health Organization [WHO]. (2022). *Health and wellbeing*. Retrived from: https://www.who.int/data/gho/data/major-themes/health-and-well-being.

Chapter 2

Roaming through Memory in Psychoanalysis (Mnestic Systems Dynamics)*

Franco Scalzone†
and Gemma Zontini

Abstract

The aim of this paper is to present the relationship which exists between the functioning of mnemic systems and human temporalities as dynamic structures/processes which might be represented as complementary images of each other.

Keywords: mnemic systems, mnemic chains, human temporalities, nachträglichkeit, recursive process

Introduction

The underlying fil rouge of the paper is represented by the relation existing between the functioning of mnemic systems and human temporalities as dynamic structures/processes which might be represented as complementary images of each other. We may also say that phantasy (innate elements), individual memory and temporality (acquired elements) function as self-referential three-fold structures able to generate a world and provide it with a meaning.

---

* This is a reformatted version of an article previously published in *Chaos and Complexity Letters*, Volume 6 Issue 1/2 (2012), New York: Nova Science Publishers.

† Corresponding Author's E-mail Address: f.scalzone@mclink.it.

In: Complexity Science in Human Change
Editors: Franco F. Orsucci and Nicoletta Sala
ISBN: 979-8-89113-408-9

The Nachträglichkeit is its main dynamic mechanism which, with its continual functioning – just like the flying shuttle of a loom – constantly corrects the warp through a temporal-spatial to-and-from movement of the desire thread.

There might therefore arise the need to individuate in the same subject different chronotypes, not only on a diachronic level but also on a synchronic one.

It is possible, by grasping the mismatch between non-congruent historical times which exist in the same chronological present, to get to envision a temporal multiversum, a time with many dimensions, all present, an intertwining of different time planes, a counterpoint of temporal unbalances in the same individual who still lives in a single and shared chronological time.

Man is a caducuos system which, through primary fantasies and his tending towards the object, encompasses, concealed in himself, a sort of unlimited time – whether projected towards the past or towards the future.

Past and Memory

Freud dealt whit memory throughout his life, because he thought that: "A psychological theory deserving any consideration must furnish an explanation of 'memory'." (Freud, 1895, p. 299). Usually two aspects of the same phenomenon are meant by the term "memory:" the mnemic function and its product: memories. As regards the topic organization of memory Freud talks about 'mnemic systems' while the actual "memory-traces" are notation modalities due to particular configurations of *facilitations* for the passage of excitement through "contact-barriers" (1895). This double use of the term "memory" might on the one side bring about some confusion, but on the other, it also allows us to use the same term to cover a wide range of phenomena, which sometimes turns out to be operatively useful.

We will see how such use is more in keeping with the reality of facts, as it is not possible to separate, so simply, the function and its functioning from contents; they are two aspects of the same phenomenon. We may, for instance, say that on "low" levels the mnestic function and the mnestic systems are the same thing; i.e., the (mnestic) function coincides with the (mnestic) organisation. The memory we are here talking about is the so-called *procedural* memory, implicit and non-symbolic. (see also Scalzone and Zontini, "Psychoanalysis in Time – Time in Psychoanalysis." in print).

Time is an organiser of mnestic systems which, in turn, organise time.

Freud connected movement to memory and therefore movement entered not only into the creation of the single memory but also into the process of the creation of it. We constantly remake the past and so memory without the present cannot exist. Furthermore, the reconstructed present originates in the past reconstructed through a *recursive process;* the past creates the present, which re-creates the past. An important difference must therefore be drawn between *sub-symbolic memory*, distributed, based upon sensory-motor schemes (entailing unconscious motor automatisms) or upon an unaware affectivity (which appears earlier in the course of phylogenesis), and *symbolic memory*, linked to thought, a highly specialised function, and to language and directly conscious; it is the ultimate product of the unremitting evolutive process of the nervous system, and it appears later.

Another issue is the one regarding the memories-recalling mechanism which is influenced by *affects*, and by the emotional context in general, which decides about their *cathexis* or de-cathexis, especially in the case of a transference situation, as in analysis. Therefore, in this case we must differentiate the mnestic apparatus and its functioning from its contents. In the *letter to Fliess of December 6, 1896*, Freud intended to demonstrate the assumption that the psychic mechanism formed following a *stratification* process. Such a process involves mnemic traces being constantly rearranged along new relations; memory, therefore, is present in a multiple form and fixated in different types of signals; moreover, every single transcription is different from the others according to the type of neurons transmitting it:

> "*W* [*Wahrnehmungen* (perceptions)] are neurones in which *perceptions* originate, to which consciousness attaches, but which in themselves retain no trace of what has happened. For *consciousness and memory are mutually exclusive*.
>
> *Wz* [*Wahrnehmungszeichen* (indication of perception)] is the first registration of the perceptions; it is quite incapable of consciousness.
>
> *Ub* [*Unbewusstsein* (unconsciousness)] is the second registration, arranged according to other, perhaps causal, relations. Ub traces would perhaps correspond to conceptual memories; equally inaccessible to consciousness.
>
> *Vb* [*Vorbewusstsein* (preconsciousness)] is the third transcription, attached to word presentation and corresponding to our official ego."
>
> Any "[...] successive registrations represent" in Freud's view "the psychic achievement of successive epochs of life" (in Masson, 1985, pp. 207-208).

Freud, in fact, says:

> "The unconscious is quite timeless. The most important as well as the strangest characteristic of psychical fixation is that all impressions are preserved, not only in the same form in which they were first received, but also in all the which have adopted in their further developments." (Freud, 1901, p. 275, note 2).

We can find useful descriptions of the organization and functioning of memory in "Studies on hysteria." where he describes the complexity of the arrangement of mnemic traces which unfold through mnemic chains. We have so far been talking about memory in order to clarify, obviously, how closely connected it is to the organization and temporal processing of psychism.

Let's quote Freud:

> "From a single scene two or more memories are reached at the same time, and from these again side-chains proceed whose individual links may once more be associatively connected with links belonging to the main chain. Indeed, a comparison with the genealogical tree of a family whose members have also, intermarried, is not at all a bad one." (Freud 1896, 198).

ORGANIZATION OF THE PATHOGENIC MATERIAL OF A SEVERE HYSTERIC NEUROSES

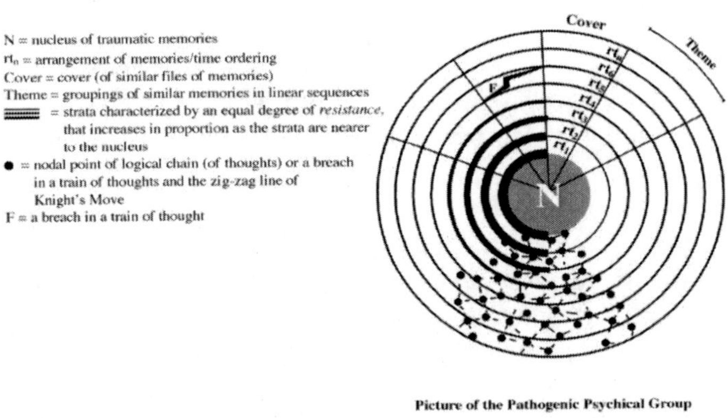

Picture of the Pathogenic Psychical Group

Figure 1.

Later on, in the last part of "The Psychotherapy of Hysteria." Freud will extensively deal with the description of the arrangement of psychic material in a "severe neuroses:" a structure in several dimensions which organizes itself in three different ways. By following the description of such a complex pathological organization of part of the mental apparatus, and precisely the repressed material, we may see, as if through a torn curtain, the structuring of part of the unconscious systems (repressed unconscious). We will use this description to show the organization of mnemic systems in a spatial-temporal sense: see figure 1.[1]

To sum up, we will say that there exists a *nucleus* consisting in memories of events or trains of thought in which the traumatic factor has culminated or the pathogenic idea has found its purest manifestation. All around such a core there is an 'atmosphere' of mnemic material arranged in a threefold order.

1) If we call *theme* the grouping of homogeneous memories into a linearly stratified plurality, the first arrangement will present itself according to a linear chronological order, like in a *dossier*; the first memory is what Freud called the *outer cover* of the dossier.
2) The second type of morphological arrangement of the themes will be the concentric mode which develops around the *central pathogenic nucleus* in strata of equal degree of resistance, though gradually increasing when approaching the centre. This type of arrangement is simultaneously morphological and chronological, because the most superficial themes are the most recent ones as well as the ones presenting less resistance. Moreover, it is an arrangement of the energetic type because resistances are provided with a countercharge regulated by its distance from the repressed: a defensive countercatexis.
3) And finally, an arrangement according to thought-content, the linkage made by a logical thread which reaches as far as the nucleus and tends to take an irregular and twisting path, different in every case: a dynamic arrangement. (Freud, 1893-1895, pp. 288-290).

This logical connection might be graphically represented with a network of lines, i.e., mnemic chains, which meet in/branch out of nodal points, thus intersecting, uniting or diverging in all directions, as if to form a web. The first type of arrangement seems to hint at an almost localistic description of fixed mnemic traces. The type of arrangement which appears to be the most

---

[1] The graph in figure 1 has been processed by Franco Scalzone and carried out by Mrs. Claudia Vignale.

interesting is the dynamic one, arranged according to logical threads. This arrangement is structured like a network, made up of *connection tracts* and *exchange nodes*, tri-dimensionally expanding around the *pathogenic nucleus* according to a spatial representation which we imagine as being spherical.

Taking into account Freud's quotation when he says: "I have therefore been obliged to put it together from even smaller fragments than are usually at one's disposal for purposes of synthesis. This task, which is not difficult in other respects, finds a natural limit when it is a question of forcing a structure which is itself in many dimensions on to the two-dimensional descriptive plane." (Freud, 1914, p. 72) and of the limits he denounces, we may imagine the graph *not* as a *bi-dimensional* structure, but as a *pluri-dimensional* structure.

Connections are established, for instance within therapy, by free associations. In this case, for the exploration of the network of the mnemic traces, at times this is resolutive "the zig-zag line in the solution of a Knight's Move" with which it is possible to enter ever-new paths Transversal and transradial, leading to key-memory, always of a sexual nature.

We may therefore access memories both through temporal sequences (*sequential access*) and at random (*random access*) as Freud said relating how one should proceed in the dream interpretation technique (1922). The organisation of the associative paths of the memory, as well as language and dream together with *condensation* and *shifting*, indicate all the mechanisms of thought movements according to the rules of its functioning.

We may then assume that memory thus organized is not allocated, within exact locations, but rather lies in a state which we might define as being "distributed". This type of allocation, or better still, of organization, is completely different from the initial two which Freud himself called "morphological". Memory is dynamically *distributed* throughout the system and therefore enjoys some peculiarities. The dynamic allocation of memories, or of the material able to organize and re-organize memories – or better still "re-categorize" within time, both spontaneously and by evocation – does not belong to an only block or neuronal map. The possibility for a memory to be evoked, therefore, also depends on the mnemonic complex of which it is part and therefore it can, for instance, be available within a certain mnemic context, within a given circuit, and conversely inaccessible within others because it can be repressed, un-cathected and therefore un-facilitated. In therapy, this will be fundamental for the retrieval of repressed memories and for the overcoming of resistances. The subject experiences itself in multiple temporal dimensions, of which only one at a time may appear within consciousness and occupy it, thus giving the impression of being the only one, and excluding the others: that particular one will be the conscious dimension of the *present* at that given moment.

Finally, the psyche must retain a recording of its own functioning and of its unfolding in time at the very moment they occur. Memory turns the past into a sort of endlessly explorable *virtual world*, and such explorations become themselves elements of memory: a second grade memory, i.e., *memory of memories*… a sort of *metaverse*.

Man is a complex system for adaptive reasons, that is to adapt himself to the complexity of the external world, of which, somehow, his own memory is part; such complexity he himself constantly tends to increase.

The graph in figure 1 visually and structurally recalls the representations of the Greek and Christian Universe, but also the ones of the *Keshvar* of the *mazdea* imaginary geography (see Corbin 1979, 47) and of Dante's cosmos, as he presented it in the *Divine Comedy* (see Patapievici 2004), or even the *mnemonic wheels* by Bruno (see Mino 2001), true *thinking machines* (*machine logica*): i.e., *alphabet-wheeled mnemonic machines*.

The significance that mnestic systems and the mnestic function have in the functioning of self-organising systems is evident, so much so that some believe that the mind functions as a "simple" mapping machine and that also a "high" level function such as intelligence may be nothing but a good management of "row memory".

To conclude, we will use the following quotation still to show other aspects of the Freudian representation of memory in relation to time. In "Civilization and its Discontents" Freud uses three metaphors to illustrate some characteristics of the "psychic organism" and stresses similarities to and differences from other systems.: 1) The metaphor of the evolution of animal species, 2) The one regarding the evolution of the "Eternal City" and 3) that one of the development of "the body of an animal or a human being". All three recall the idea of an evolution, though they are lacking, as he himself tells us "[...] that only in the mind is such a preservation of all the earlier stages alongside of the final form possible, and that we are not in a position to represent this phenomenon in pictorial terms." (1930, 71).

For reasons of space we need to limit ourselves and only consider the "Eternal City evolution" metaphor.

Let us listen to Freud: But have we a right to assume the survival of something that was originally there, alongside of what was later derived from it? Undoubtedly. There is nothing strange in such a phenomenon, whether in the mental field or elsewhere. […]

This brings us to the more general problem of preservation in the sphere of the mind. The subject has hardly been studied as yet;1 but it is so attractive and important that we may be allowed to turn our attention to it for a little, even though our excuse is insufficient. Since we overcame the error of supposing that the forgetting we are familiar with signified a destruction of the memory-trace—that is, its annihilation—we have been inclined to take the opposite view, that in mental life nothing which has once been formed can perish—that everything is somehow preserved and that in suitable circumstances (when, for instance, regression goes back far enough) it can once more be brought to light. Let us try to grasp what this assumption involves by taking an analogy from another field. We will choose as an example the history of the Eternal City. [...]

Now let us, by a flight of imagination, suppose that Rome is not a human habitation but a psychical entity with a similarly long and copious past—an entity, that is to say, in which nothing that has once come into existence will have passed away and all the earlier phases of development continue to exist alongside the latest one. This would mean that in Rome the palaces of the Caesars and the Septizonium of Septimius Severus would still be rising to their old height on the Palatine and that the castle of S. Angelo would still be carrying on its battlements the beautiful statues which graced it until the siege by the Goths, and so on. But more than this. In the place occupied by the Palazzo Caffarelli would once more stand—without the Palazzo having to be removed—the Temple of Jupiter Gapitolinus; and this not only in its latest shape, as the Romans of the Empire saw it, but also in its earliest one, when it still showed Etruscan forms and was ornamented with terracotta antefixes. Where the Coliseum now stands we could at the same time admire Nero's vanished Golden House. On the Piazza of the Pantheon we should find not only the Pantheon of to-day, as it was bequeathed to us by Hadrian, but, on the same site, the original edifice erected by Agrippa; indeed, the same piece of ground would be supporting the church of Santa Maria sopra Minerva and the ancient temple over which it was built. And the observer would perhaps only have to change the direction of his glance or his position in order to call up the one view or the other.

There is clearly no point in spinning our phantasy any further, for it leads to things that are unimaginable and even absurd. If we want to represent historical sequence in spatial terms we can only do it by juxtaposition in space: the same space cannot have two different contents. Our attempt seems to be an idle game. It has only one justification. It shows us how far we are from mastering the characteristics of mental life by representing them in pictorial terms. [...]

The fact remains that only in the mind is such a preservation of all the earlier stages alongside of the final form possible, and that we are not in a position to represent this phenomenon in pictorial terms. (Freud, 1930 [1929], pp. 68-71).

We wish to underline the difficulty Freud finds here, which is being able to represent in spatial terms, and therefore visual, the simultaneous presence of mnemic traces belonging to different historical times: different *chronotypes* (see Fachinelli 1979). Such a situation may be better illustrated today by using a spatial representation based upon the concept of memory *maps* or transitory *activation patterns* or *circuits*. Retrieving a memory entails the activation of some portions of global maps. In the same *neuronal map* there are various possibilities of carving out different *neuronal groups*, which once activated, give rise to different memories. The same 'ψ neuron' present in a map – neuron here is meant as a *virtual* structure and not an anatomical one – can join different activation *patterns*, and therefore different memories, according to the pattern which has been activated there and then. This is why if we imagine ancient Rome as a unique map, though provided with numerous and different activation patterns, we may spatially picture what was already stated by the Freudian model itself. As a matter of fact, within the same map there coexist many transitory patterns, all on stand by, ready to be activated and capable of giving rise to a memory, without the single memory itself being stored as an engram or fixed image somewhere into the cerebral tissue. Thus indeed:

> [...] the observer would perhaps only have to change the direction of his glance or his position in order to call up the one view or the other." (see above) and, by changing the activated pattern, the outcome would be that within psyche, the same space might have two different contents. There can be endless states of memory, overlapping and coexisting, without there being any *overprinting,* and therefore numberless configurations. The brain truly appears to be an "magic loom"

We must now add to this description of the mnestic systems organisation and of memories, already in themselves complex, a further element of complexity represented by the existence of a particular kind of memories: *screen memories* which somehow falsify our memory. They can be of three types of "chronological relation between the screen memory and the content which is screened off by it." (Freud, 1901, p. 44): There are different types of *screen memories*: *retro-active* or *retrogressive, pushed ahead* or *displaced forward* e *contemporary* or *contiguous*.

However, the memory map can be modified in certain points due to resistances, therefore "[...] instead of the mnemic image which would have been justified by the original event, another is produced which has been to some degree associatively displaced from the former one." (1899, p. 307). The shifting is of a spatial-temporal nature. As Freud stresses, the process, which occurs in the psychic life both of psycho-neurotic subjects, because of a *conflict*, and in normal subjects is:

conflict → repression → substitution involving a compromise

In such cases the modification of the spatial-temporal representation of memory is due to the fact that, in such occurrences, repressed fantasies tends to pass into a mnestic trace of childhood scene which has some points of contact with the former, which leads to a falsification of memory. One last but very important thing is that: "[...] since what provides the intermediate step between a screen memory and what it conceals is likely to be a verbal expression (Freud, 1899, p. 319).

> "But in these cases too the falsified memory is the first that we become aware of: the raw material of memory-traces out of which it was forged remains unknown to us in its original form." (Freud, *ibidem*, p. 322).

In the case of memory, Freud does not always put forward the same ideas. Sometimes he appears to be talking about memory as static, *localized*, based on fixed memory traces which, when activated, produce memories; at other times he presents the picture of a dynamic memory, arranged in a *distributed* way and capable of constant re-categorizations. Actually, in our brain there are no specific memories; there are only tools for re-organizing past impressions. A memory of such a type is both associative and inaccurate – *reminiscences* – unlike the replicative memory of computers, but it is dynamic and capable of generalizations, because founded upon probabilistic perceptive categorizations, in turn based upon "values." and constant re-categorizations (see Edelman, 1992). Freud wonders:

> "What then is the physiological correlate of the simple idea emerge and re-emerging? Obviously nothing static, but something in the nature of a process. This process is not incompatible with localization. It stats at a

> specific point in the cortex and from there spreads over the whole cortex and along certain pathways. When this event has taken place it leaves behind a modification, with the possibility of a memory, in the part of cortex affected. [...] Yet whenever the same cortical state is elicited again, the previous psychic event re-emerges as a memory. We have, of course, not the slightest idea how animal tissue can possibly undergo, and differentiate, so many various modifications." (Freud, 1891, p. 56).

If we now piece together both quotations, the one relating nucleuses of pathogenic matter and the one concerning the analogy with the Eternal City, we see that Freud shows us a very complex organization of memory in which, to the three spatial dimensions, the dimension of time is added. For this reason, it will be impossible for him "to represent historical sequence in spatial terms" but it will not be unimaginable. It is true that it is not possible to portray it, as "the same space cannot have two different contents." but only if we use a static representation. If, instead, we use a dynamic representation through moving images, then we can see all the various eras flowing on a screen without the images of the past being erased by images of the present: we only need to underline and fix the images which at a certain time constitute the present. We can control anything but the past and the one of all those who are in our "cone of past".

Just a few remarks about *phylogenetic memory*. The idea is as early as "Studies on Hysteria". In "Totem and Taboo" (1913-1914), Freud hypothesises "the existence of a collective mind, in which mental processes occur just as they do in the mind of an individual." (p. 157). Probably he made such an hypothesis in the attempt to search for an *hidden variable* which might function also as an unknown determinant of the consequences of fantasised traumatic events. [*primary phantasies*].

After the formulation of the second topic, the phylogenetic memory found its well-defined collocation within the psychic apparatus and it acquired a fundamental function in the development of an individual. The id became the depositary of the archaic heritage (inheritance) of individuals.

We would like to conclude by noticing that in our view, the interminability of analysis results from the fact that investigation of the unconscious in general, and therefore also of dreams, generates its object at the same time as it explores it: We, then, find ourselves in the presence of a *transfinite*. This process activates a *multiplier of meanings*, so that the reservoir of meanings does not shrink with the progress of interpretative activity. Instead, the boundary between the analysed and the analysable 'shifts constantly forward', resulting in a Zuider Zee that can never be completely drained. Moreover, every interpretation throws light on a new part of the dream to be interpreted – but, to make the work of interpretation inexhaustible, it is not necessary for this part of the dream to be uninterpretable, but only for it to be interpreted in such a way as to generate new parts to be interpreted in turn, and so on in an interminable process (see Scalzone e Zontini, 2001), which is not human life, that is submitted to the laws of open, dissipative systems; therefore it has a natural ending.

Also here we may notice the presence of an *interminable* inscribed into a *finite* because we must observe, as Prigogine said (1988), that to the illusion-time of Einstein's *irreversibility* and to the *degradation* time of enthropy we must add the creative time: the *constructive* time of complex systems.

Regarding this issue Gaddini writes:

> "There remains one more significant factor regarding self-loss anxiety to which we should give our attention: the fact that, with its insurgence, the mental processes are introduced to the dimension of time. [...] As distinct, that is, from the mind's use of the memory in the phase of non-integration, self-loss anxiety continually augments the memory of the catastrophe as a past event which must remain in the past, a 'before' which must never become 'now', but which rather institutes a 'past'. What I mean is that for the first time the memory becomes 'temporalized'. However, if detachment has been traumatic, this temporalizing function of self-loss anxiety may not take effect. In this case the memory maintains its actualizing activity, so that it always seems the catastrophic experience may happen 'now'. The pathological result is the continual expectation of the catastrophe." (see Gaddini, 1980, p. 140).

The birth of memory, as an irreversible phenomenon, breaks up the symmetry and generates the birth of time. Therefore, also our time is a time constructive of new structures like the mnestic ones and.... *vice versa*.

References

Corbin H. (1979). *Corpo spirituale e corpo celeste*. Milano, Adelphi, 1986.

Edelman G.M. (1992). *Bright Air, Brilliant Fire. On the Matter of the Mind*, London, Penguin Books.

Fachinelli E. (1979). *La freccia ferma* [*The still arrow*]. Milano, L'erba voglio.

Freud S. (1891). *On aphasia: a critical study*. New York, IUP, 1953.
Freud S. Breuer J. (1893-1895). *Studies on Hysteria*, SE 2, p. 143–56.
Freud S. (1895). *Project for a scientific psychology*. SE 1.
Feud S. (1899). *Screen Memories*. SE 3, 301-322.
Freud S (1900). *The interpretation of dreams*. SE 4-5.
Freud S. (1901). *Psychopathology of everyday life*. SE 6.
Freud S. (1913-1914). *Totem and taboo*. SE 13.
Freud S. (1913-1914). *From the History of an Infantile Neurosis*. SE 17.
Freud S. (1923 [1922]). *Remarks on the theory and practice of dream-interpretation*. SE, 19.
Freud S. (1930). *Civilization and its discontents*. SE 21, pp. 57-146.
Gaddini E. (1980). "Notes on the Mind-Body Question". In Limentani A. (edt.), *A Psychoanalytic Theory of Infantile Experience*, pp. 119-241, London, Brunner-Routledge, 1992.
Masson J.M. (edt.) (1985). *The Complete Letters of Sigmund Freud to Wilhelm Fliess 1887-1904*. Cambridge Massaciusset and London England, Harvard University Press.
Patapievici H.-R. (2004). *Gli occhi di Beatrice*. Milano, Bruno Mondadori, 2006.
Prigogine I. (1988). *La nascita del tempo*. Roma-Napoli, Theoria.
Scalzone F., Zontini G. (2001). "The Dream's Navel between Chaos and Thought". *International Journal of Psychoanalysis*, 2001, 82, pp. 263-282.
Scalzone F., Zontini G. *"Psychoanalysis in Time – Time in Psychoanalysis"*. (In print).

Chapter 3

Reconstructing Experience in Personality Organization with Physical or Semantic Reciprocity*

Bernardo Nardi†

Department of Clinical and Experimental Medicine, Polytechnic University of Marche, Ancona, Italy

Abstract

In my model of Personality Organizations (POs), the assimilation and reprocessing of experience according to the route of development of individual adaptive competences allows focusing on the physiological resources provided by each subject's PO. In fact, each PO allows developing some *specific sensory-perceptual susceptibilities*, based on the fact that it appears more "*permeable*" and "*sensitive*" to the more significant aspects and to the *predominant domains* in its environment. In this way each PO evolves some distinctive assessment and operational strategies, and achieves selective and specialized adaptive resources that are then honed in the course of life. The reformulation of experience can help identify and bring forth the potential resources and adaptive competences – inhibited at the tacit level and ignored at the explicit level – that are related to the PO of the subject undergoing therapy. It is possible to see how on the one hand there is a constancy related to the consistency of development processes that lead to a specific basic organizational closure (Controlling, Detached, Contextualized or Normative types); on the other hand, it is possible to observe the complex

---

* This is a reformatted version of an article previously published in *Chaos and Complexity Letters*, Volume 10 Issue 2 (2016), New York: Nova Science Publishers.

† Corresponding Author's E-mail Address: bernardonardi@outlook.it.

In: Complexity Science in Human Change
Editors: Franco F. Orsucci and Nicoletta Sala
ISBN: 979-8-89113-408-9

evolutionary dynamics connected with the adaptive abilities used to assimilate and self-refer experience. The reformulation of experience thus provides access to new ways of living it that are consistent with the subject's PO, thus also changing the matching behavioural expressions and highlighting those subjective resources that make each individual unique.

Keywords: adaptive response, evolutionary stress, physical/semantic reciprocity, Personality Organizations (POs), post-rationalist psychotherapy

1. Subjective Modalities of Assimilating Experience and Sense of Self

The reformulation of subjective experience is at the heart of the post-rationalist psychotherapeutic process, whose guidelines have been laid down by Vittorio Guidano (1987-2010) in his model of Personal Meaning Organizations (PMOs).

The PMO model was inspired by psychopathology, with the normal functioning of each PMO being described by Guidano starting from its most common clinical manifestations. Over the past several years I have followed an inverse route: I have started from the adaptive resources provided by each Personality Organization (PO), which I consider as the maximum expression of the adaptive ability of the complex system constituted by the human psyche. This is why I have given each PO a "physiological" name and have sought its origin in a specific adaptive response to a specific evolutionary stress. Based on these considerations I have sought to identify the resources that are at least potentially provided by each PO also in individuals with current psychopathological disturbances, and have considered such resources as the fundamental levers on which to act to induce a stable adaptive change (Nardi, 2007-2016).

Cognitive research has documented that subjects undergoing psychotherapy tend to present their problem in an objective form, like something from which they are suffering, that appears foreign and external to the self, and which destabilizes and distresses them. On the basis of this life experience, the approach of focusing on a recent episode, where the subject has experienced the problem at issue, has not only the potential to highlight the life theme (e.g., control, adequacy, personal perfection) underlying the problem, but also to enable exploration by the subject of his/her subjective mode of assimilating experience, to use in the construction of his/her identity. To focus on his/her way of functioning makes it possible for the subject to seek novel ways to achieve a sense of self that are more adaptive and flexible than the earlier ones, and that are consistent with his/her own PO.

Of the two components of subjective experience – the immediate (direct) and the explained (mediated by language) – it is the former that provides the rough material that is required to assimilate the things one lives through and to integrate them into the sense of self. Indeed, immediate experience provides direct information on what the subject feels (e.g., simple and complex sensations, feelings, perceptions, and mental representations); how he/she does so (in terms of emotional activations); and when (i.e., the time context). Focusing on immediate experience also enables the subject to rebuild the corresponding level of subjective awareness, which is based on implicit, "tacit" knowledge. This basic knowledge level is partially accessible by awareness, it is general, global (holistic), strongly affected by emotional activations, and provides a single, continuous representation of the self and the world through automatic and schematic (procedural) forms of learning and memorization. At the tacit level, conflicting experiences that occur with a certain regularity lead to the formation of an internal general (cenesthetic) feeling of the self.

Therefore, a subject's ability to assimilate experience depends on how and to what extent he/she is able to self-refer it, and this requires the experience to be sufficiently compatible with the tacit subjective modalities adopted until then.

As can be noted from the arising of resistance to change, the wellbeing induced by a new experience does not in itself make it capable of being assimilated, whereas new and even negative experiences compatible with the tacit representations (and relevant emotional activations) through which subjects have learned to recognize themselves can be assimilated.

From the tenderest age, perceived experience – which consists of a set of internal representations connected with a variably wide range of emotional activations – forms in a discontinuous manner, in association with "typical situations" that induce emotional activation in the child. The accretion of such situations and of their representations and internal activations leads to consolidation not only of memories of the experiences, but also and especially of the activations produced by them ("the effect that being the one who made that specific experience had on the subject"). This contributes to the construction of the self to an extent that varies according to the strength of the perceived activation. Thus, in current experience the subject refers to the past through memory and expects the future through imagination. As subjectively significant episodes accumulate, early self-perceptions gradually strengthen, they become incorporated into a continuous flow of representations (together with the relevant situational emotional activations), and concur to form an early unitary sense of self. All subsequent experience is compared against it and found to be more or less apt to be assimilated based on its difference from the sense of self that has matured until then.

Another critical aspect, that should be underlined, is that the significant emotional activations in which subjects recognize themselves gradually acquire a predictive value, anticipating what subjects begin to expect from a new experience; for instance, an activation that is predominantly of loneliness generates the expectation that when a new situation arises the subject will be able face it with his/her own resources and will not need external help. For instance, a feeling of being loved and accepted leads to the search for further positive confirmations in new scenarios, a behaviour that applies both to predominantly favourable feelings and, unfortunately, also to predominantly negative ones. This is the territory of the psychotherapist, who supplies the tools that can be employed by the subject to discover his/her personal resources, both present and potential, irrespective of the feelings highlighted by life experiences.

The above considerations explain how at the tacit level the subject seeks to reproduce new situations which are similar to significant earlier situations that are felt to characterize one's life; this entails a confirmation of what the subject feels he/she is, whereas at the explicit level the subject does not seem to be aware of having actively sought the situations (for instance: "you see, this is the way I am, this is my destiny, there's nothing I can do to change it…"). In addition, in such cases it is easy to observe that the facets of experience that are not consistent with the sense of self in which the subject recognizes him/herself tend to be perceived as unimportant and scarcely relevant, and set aside, waiting for something that will confirm the expectations in which he/she recognizes him/herself (a subject whose sense of self is built on a negative register will say: "yes, this period is going fairly well, but you'll see that sooner or later something will go wrong and will ruin everything again"). In other words, at the tacit level only the life experiences connected with the episodes or existential issues that the subject has learnt to recognize as characterizing his/her life – producing emotional states that are comparable to those that have consolidated in the emotional memory at the tacit level – are considered, whereas all those episodes that are not perceived as matching the contents of the internal representations are considered unimportant, marginal, occasional. It is a situation reminiscent of the game where one is required to draw lines between some dots in a frame, ignoring all the other dots, to obtain a predetermined figure: the game does not envisage linking all the dots in different ways, as this would provide a large number of potentially alternative images.

This human modality of assimilating experience to build one's identity – at first sensory-perceptual and emotional and later also cognitive – is critical not only to focus on the tacit individual processes related to the construction and maintenance of the sense of self, but also to understand the origin and nature of the resistance to change that arises.

Therefore, reformulating experience during psychotherapy enables the subject to begin focusing on the fact that the modalities he/she has always used to self-refer experience are merely the habitual modes by which he/she has learnt to live, whereas other modalities that are adaptively more advantageous can be employed and recognized as their own, thus enabling incorporation of the new perceived emotions into the sense of self. From this viewpoint, the psychopathological symptoms, which reflect an attempt to preserve internal consistency with respect to the world, can also be a starting point to explore personal functioning. The subject thus discovers further possibilities besides the approach – hitherto considered as ordinary and natural – of tying personal experience and what one is, including one's potential resources, indissolubly to one's history, and will discover that history is merely events that have happened to him/her, which therefore have no direct connection with one's quality, worth, and are even less related to one's destiny.

The research into the adaptive capacity to assimilate experience, that I have conducted over the past few years (Nardi 2007-2015), has clearly indicated that POs are the most complex modalities of reading experience of human mind, and that they allow balancing, in a highly specialized, subjective way, the opposite needs for continuity and change that arise in the course of life. In other words, along the evolutionary journey towards Homo sapiens, the improved operational capacities of the central nervous system have led to the diversification of some modalities, that have become stable in each individual, whereby experience is read based on a number of specific aspects that are exploited not only to evolve a behavioural specialization, but also to define one's identity at the tacit and the explicit level. In fact, each PO enables recognizing and learning to manage, in the relational and physical environment where the individual grows up, some fundamental variables directly related to survival – that are far more selected and evolved that those shared by other species – leading to the development of highly evolved individual competences that confer benefits both at the personal and the relational level.

2. Adaptive Modalities of Self-Referring Experience in Personality Organizations

Any experience that the subject lives through as new and unexpected is a stressor for his/her ability to adapt, and produces an internal activation at multiple levels, including the neurovegetative, musculoskeletal, neuro-immune and, especially, the emotional and cognitive levels.

A reliable caregiver, who appears to the child as a safe refuge, is capable not only to express that *empathy* that makes the child feel welcomed, reassured, and therefore encouraged to relate to the environment, but also knows how to *regulate the child's exposure* to environmental stressors in relation to the competences and skills he/she gradually develops. In addition, such caregiver knows how to *valorize* the child's unique aspects, making the comparison with others an opportunity to discover their resources and to share, where this is possible, some aspects of experience, resulting in mutual enrichment.

A. The Experience of POs with Physical Reciprocity (PR)

The consistent repetition of the caregiver's non-verbal behaviour in the same situations – in the form of similar internal activations expressed by the same basic emotions – soon turns stressors into "*behaviours that can be expected*" by the child, who learns early tacit recognition of the internal states he/she uses to self-refer current experience.

Since the internal activations are initially the same as those of the caregiver, the child very soon stops feeling the need to refer to the caregiver to understand whether the situation he/she is experiencing is safe. Subsequently however, the acquisition and maturation of new operational competences involves discovery of his/her ability to manage a given stressor (e.g., a danger situation); this entails a change in his/her internal state, related to the awareness of having now the ability to control a situation that previously appeared as dangerous, and in the relevant emotional activations (for instance, the child feels safe and is no longer afraid). In subsequent physiological development, the changes in the internal state begin to become uncoupled from the caregiver's external behaviour. For instance, a parent may continue to fear for the child's safety in a situation where the child has achieved the competence required to manage alone; in such situation the child can read the caregiver's anxiety simply as a demonstration of the parent's love for him/her.

In POs with *Physical Reciprocity* (PR), relationships are predominantly managed in *physical and spatial terms* through approaches and retreats with respect to the figures and the contexts that are significant for the subject, in the way that best matches his/her internal states.

The reading of experience is therefore based primarily on the internal sensory-perceptual activations and the relevant basic emotions, with an early focusing on internal signals, which become central and guide the child in the definition of the external context. Experiencing internal wellbeing or *discomfort* thus induces *regulation of tacit self-knowledge* in relation to the context and a consequent (possible) behavioural adjustment to the situation the child is facing.

The assimilation of experience – which becomes incorporated into the sense of self that the subject has developed until that moment, in a continuum between positive and negative sense of self that defines personal realization – is guided by *internal sensory-perceptual activations* and by *basic emotions* in relation to the two key *stress categories* that the subject learns to recognize, control, and manage to achieve a sense of personal realization: *danger* (in relation to all that can constitute a threat), and *loneliness* (as in situations where it is necessary to act in first person).

In the PRPO development routes, the reading of internal signals develops in infancy, and is centred on the management of one's wellbeing or discomfort to regulate the relationships with the external environment; this enables the child to develop a growing ability to set the tacit focus on the management of danger or loneliness situations.

The discovery and valorization of one's abilities and common sense enable the practical organization of one's activities, and through them the development of relationships, which however need to pass the test of reliability (figures on whom one can count).

The reorganization that takes place in adolescence, characterized by maturation of abstract and self-reflective thought, involves a change in the use of internal signals. When a good mentalization capacity is present, the affective relationships with family and friends and the discovery of new, significant relationships can become more adequate, profound, and flexible through introspection and the utilization of symbolic functions, which enable a more specific and selective reading of internal states. Specific adaptive competences are thus developed to manage the situations and problems that need to be addressed to achieve self-realization, making reference to reliable figures and harnessing the personal competences that have gradually been developed.

Equilibrium is achieved through the ability to manage the subjective danger signals; the subject achieves an increasingly mature and complex, tacit and explicit knowledge of wellbeing and personal equilibrium. This occurs in PRPO development pathways, both those where the environment appears "*highly available*" and "*accessible*" (e.g., *Controller* PRPOs) and those where it appears "*scarcely available*" and "*poorly accessible*" (e.g., *Detached* PRPOs).

A1. The "Online" Experience of Controller PRPOs

In a PRPO environment - where the care-giver appears close and available with respect to one's internal needs, providing warmth and a sense of protection ("*highly available*" and "*accessible*" care-giving) both over time, situation after situation, each time one feels the need for it ("*on-line*" care-giving), and in space, ensuring prompt assistance ("*tight rein*" care-giving) - there develops an ability to identify the dangers that threaten the control of one's life as well as the protective reference figures experienced as reliable and the instruments required to manage situations.

High, tight rein, on-line availability allows one to use, when this is helpful, the available reference figures to learn to be operational and solve problems, to get confirmations of who one can count on, to strengthen significant relationships, and to exchange skill and abilities.

The stressor represented by *danger* (a direct or indirect, concrete and simple, abstract and complex threat) thus leads to the development as a fundamental competence of the ability to manage the various situations so as to live them safely. Such managing ability is obtained from an awareness of the internal states, particularly the physical and emotional tones of peacefulness or anxiety, which are closely related to currently available operational tools.

The subject gradually discovers that he/she can not only perceive, but also identify reliable reference figures in relation to the need to feel safe, while at the same time enjoying areas of freedom. Such tacit attitude prompts the subject to manage life contexts personally or, when this cannot be done, to identify those who can do it reliably. The reliability of others is established precisely from the response the subject receives with respect to his/her internal needs, and this is the discriminating factor guiding in the identification of significant references and in the construction of the relational network.

The reading of experience by Controller PRPOs thus passes through the development of three key *adaptive skills*: a) to learn to pre-empt and manage dangers promptly; b) to identify new reliable figures (as necessary); c) to find the most effective ways and means to manage situations. As a consequence Controller PRPOs, when they face experience, tend to develop two *abilities* that confer considerable personal and social adaptation advantages: "*alertness,*" which enables prompt identification of what needs to be done, also in view of the possible consequences of what is happening, and "*identification of those on whom one can count,*" by selecting significant persons and ways and means that are considered reliable, while discarding those figures and tools that appear to be neither reliable nor useful. The ability to seek reliable reference figures and available spaces to manage difficulties and dangers thus reflects the tacit resource that the subject may discover to possess and that he/she can begin to use (or use again).

In the *development routes* of Controller PRPOs, the reading of experience – in the immediacy of its subjective perceptions, images, and feelings – involves its decoding in terms of safety/danger; this is the basis on which the subject takes control of the situation and assesses how reliable is the context in which he/she finds him/herself. The need to identify figures and environmental contexts perceived as reliable and to remain close to them, initially in physical terms but later in abstract terms, is the precondition that allows developing his/her life projects and fine-tuning the balance between attachment and exploratory behaviour, which remain interdependent. The feelings of safety and self-confidence fostered by the emotional warmth received from the reference figures allow a gradual extension of the subject's range of action, exploration of unknown environments, and acquisition of new competences and self-confidence, making him/her a reliable reference for others.

During *adolescent maturation*, the emergence of abstract thought allows interiorization of the immediate experience of safety/danger that is felt in relation to the profound changes in physical efficiency or the relational scenarios that are typical of this stage of life. The basic emotional tones, particularly fear, are integrated and managed by more complex and diverse cognitive contents, to enable the search for new projects and activities (related to friendships, work, sports, music, etc.).

When the need arises, the subject may undertake affective and *relational investments* – either early on or in more mature stages – that provide protection and freedom, and put them to tacit test. At other times the search for new horizons to express their operational skills prompts subjects to avoid stable and early affective investments, or to distance themselves from the family environment. The family however remains the reference even though the subject lives far away or abroad, and to the family the subject will eventually return. In all cases significant relational investments are tacitly guided by the search for and the proof of a reliability and availability that match one's internal states. Clearly, the *bonds* that are perceived as being more *reliable* (at the affective if not at the operational level) are still those with *family*. Reliability and availability with respect to perceived internal needs guide in the selection and maintenance of friends on whom one can count. Similarly, when the need is felt to construct an affective relationship, the couple alliance is put to the test and then tacitly built after proof has been obtained that the partner has the wellbeing of the subject at heart, as garnered from the fact that he/she is available and active when needed, but also accepts the subject's need for his/her space.

The ability to assimilate novelties safely and to share discoveries and achievements with significant figures matures progressively in physiological development routes, moving from a predominantly physical reciprocity to an emotionally more mature relational style, one that is more flexible with regard to the opinions and life choices that can be shared with mutual advantage.

In clinical practice, in *Controller PRPOs* the reformulation of experience based on the significant episodes referred by the subject and analyzed "in slow motion" during psychotherapy, allows distinguishing the habitual way (experienced as the objective and only possible way) to self-refer experience from other possible ways that are consistent with the subject's PO and are objectively more flexible and advantageous. This also allows harnessing the resources of the subject's PO, hitherto ignored or not used, for instance through the realization of one's ability to identify reliable reference figures and room for freedom to be used to meet one's needs, or of the ability to use one's common sense to identify and manage risks and dangers as safely as possible, separating what is worth controlling from what is not useful to manage and control.

A2. The "Offline" Experience of Detached PRPOs

On the other PR side - in an environment where the care-giver appears detached and poorly available with respect to one's internal needs, providing scant reciprocity in terms of warmth and communication, thus fostering early responsibilization and maturation ("*scarcely available*" and "*poorly accessible*" care-giving), both over time, through interventions that are enacted only when they appear necessary, irrespective of demands ("*off-line*" care-giving), and in space, through remote vigilance ("*loose rein*" care-giving) - there develops an ability to learn to feel safe also when one is by oneself and to face situations *in first person.*

Poor, loose rein, off-line availability enables the gradual discovery of one's autonomy skills and their use to support one's personal commitment according to the instructions that have been received, and the execution of one's tasks. The ability to do things by oneself allows forming discrete relationships, where sharing with others however respects the spaces and ways to which one has become accustomed while alone.

Indeed, the caregiver does not get involved each time, but provides the instructions and the prompts that, keeping into account the abilities the child has developed, enable the gradual extension of his/her range of action and autonomous functioning; this allows the child to learn to take on new tasks alone, but without feeling abandoned, and gradually to acquire confidence in his/her own abilities.

The stressor represented by *loneliness* (a situation where the subject must act in first person which however also provides a space for personal experience and operation) leads to the development of the fundamental skill of self-management. The subject evolves an early sense of responsibility that prompts personal commitment. Feeling encouraged to take on experience increasingly alone enables the subject to develop early management competences, to become responsible, to acquire confidence in his/her own means, and to learn to appreciate his/her autonomy, self-determination, reliability, and responsibilization. These abilities also provide the means to weave his/her relational network and to seek new significant figures.

In Detached POs, the experience is therefore read through the development of three distinctive, key *adaptive skills*: a) to learn to face new experiences alone; b) to identify (when necessary) the ways to valorize one's autonomy; and c) to develop one's operational skills with determination, according to the inclinations highlighted by introspection. As a result, Detached PMOs tend to develop two *abilities* of managing experience that confer considerable personal and social adaptation advantages: "*autonomy in choices,*" which allows – when this is possible – to make decisions and adopt projects using their own resources without support from others, but identifying when necessary stable reference figures and discriminating them from non-stable ones, and "*long-term practical planning,*" based on the determination to complete what they have begun and using their own forces and resources.

In the *developmental routes* of Detached PRPOs – where the predominant aspect is the focus from inside a situation, a context, or a relationship, with an emotional register centred on basic emotions – the subject's ability to respond to the physical stress induced by situations requiring his/her involvement in first person allows harnessing internal activations and body signals to fine-tune the relationship with the environment, so as to keep his/her tacit self-awareness stable. Starting from the *bond of separation and detachment*, which is negative only in appearance and produces a sense of partial and never final accessibility and manageability of reality, there emerge adaptive modalities that enable devising autonomous and creative strategies and pursuing objectives that once achieved involve setting farther goals.

During *adolescent maturation*, the attempt to overcome instability and detachment, which are often perceived in this phase of life, frequently in relation to profound perceived changes, prompts the subject to invest in first person both in education, in view of future professional and work activities, and in affective relationships. Especially in the latter field, the control of sadness activations prompts testing and verification of one's investments, focusing on the solidarity that can be shared with the other even though the reality appears unstable, short-lived, and scarcely open to empathy. In this way, along the adaptive routes that begin in adolescence, it is possible to focus on life themes where, through one's involvement and practical operational skills, it is possible to devise and pursue original and personalized projects in which to believe and for which it is worth committing oneself despite the awareness of the limitations, relativity, and risks of life. The experiences that have been made thus make it possible to harness responsibility and autonomy as useful resources not only in situations that involve acting alone, but also in cases where it is possible to share something important with those who can and want to take part in the subject's social network.

The significant *relational investments* are tacitly guided by the search for and proof of reliability, responsibility, and consistency much more than of emotional warmth. The couple alliance is tested and tacitly built on the demonstration that the partner is consistently reliable, is involved with the subject, accepts him/her without asking him/her to change, and that this attitude is lasting, not contingent or transient.

In clinical practice, in *Detached PRPOs* the reformulation of experience based on the significant episodes referred by the subject and analyzed in slow motion during psychotherapy, allows distinguishing the habitual way (experienced as the objective and only possible way) to self-refer experience from other possible ways that are consistent with the subject's PMO and are objectively more flexible and advantageous. This also allows harnessing the resources of the subject's PMO, hitherto ignored or not used, for instance to discover one's ability to be autonomous and to manage situations as a useful resource not only in the absence of available reference figures, but also as a basis to share one's abilities with others. The awareness of one's autonomous management skills and attitude to carry out one's life plans also opens the possibility of being with another person in a responsible and constructive way and of starting a stable affective relationship.

B. The Experience of POs with Semantic Reciprocity (SR)

When non-verbal caregiver behaviours appear to vary in relation to the external context, or to depend on the respect of some rules more than on the child's internal needs and signals, the caregiver's attitudes at first appear to the child as "*poorly predictable*" and induce him/her to pay tacit attention primarily to the external environment, which is therefore employed early to modulate and decode internal states when self-referring experience.

The child tends to imitate the caregiver's attitudes towards the external environment, and it is precisely the priority given to external signals that prompts the child to continue to make reference to the caregiver (and later to other reference figures) to judge whether his/her own behaviour with respect to the current situation is correct. In such cases, the empathy and coherence of these significant figures appear to be critical to orient the child's affective state as he/she explores the environment and relates to others. Subsequently, through learning and the maturation of new cognitive competences, the subject begins to realize that he/she can manage some stressors (e.g., a judgement/opinion, carrying out a certain task, or achieving a given goal). The discovery induces a change in the internal state (due to having learned to manage a situation for which he/she had hitherto been dependent on the caregiver) and in the relevant emotional activations (e.g., the subject experiences satisfaction and pride, no longer worry, shame or guilt for being unable to do something).

As Guidano (1987) noted, recognizing the emotional state of a significant figure is a precondition for its decoding when that state is experienced in person; however, the subject also needs to learn to distinguish the self from the source of identification. Only in this way, by maintaining a dynamic equilibrium between the tendency towards the outside and the one towards the inside, can his/her identity be adaptively constructed. Similarly, during maturation it becomes possible to learn to select and to assess in an increasingly flexible, abstract, and critical way the confirmations and disconfirmations received from the environment, so as to undertake life projects suitable to one's internal needs.

In SRPOs relationships are predominantly modulated on the basis of a *semantic register*, based on what in the environment is perceived as important to oneself and most suitable to meet one's needs.

The assimilation of experience – which becomes incorporated into the sense of self that the subject has developed until then, in a continuum between positive and negative sense of self that defines personal realization – is therefore guided by external signals and the relevant *emotions* (*predominantly self-evaluative and social*) in relation to the two key *stress categories* that the subject learns to recognize, control, and manage to achieve a sense of personal realization: *judgement* (i.e., external attitudes that are useful to understand something of oneself and of reality through the sensory-perceptual and emotional activations it induces) and *duty* (with respect to what is indicated as being right to do).

A *semantic decoding* register of external demands develops early, with the consequent tacit focus on the environmental context induced by the motivation provided by the caregiver's attitudes. These attitudes appear at first unpredictable, and often do not march the internal needs, since they vary in relation to context requirements or norms that need to be respected; however, in the light of actual situations, expectations and external rules they gradually begin to be understood and to appear to be, at least potentially, manageable. This happens both in the case of contingent situational *judgements*, which provide the measure of what one expects "here and now," and in the case of basic values that must be respected regardless of changing situations and circumstances.

Therefore, in physiological SRPO development routes, the reading of external signals is refined early in infancy, and the internal state of wellbeing or discomfort is regulated through its match with external signals, perceived as positive or negative. This allows the subject to mature a growing ability to set the internal, tacit focus based on the match with external demands and rules, identified as significant references to manage life situations and contexts. These changes of internal states, modulated by external confirmations, progressively lead the subject towards autonomy through the identification of further significant references. In the same way, the relationship with family caregivers appears to be increasingly modulated by the *quality of feelings and mutual esteem*. When this does not happen, the insecurity activations tend to induce a *closure* with respect to the environment on the passive side, and a *challenge* based on opposition and transgression on the active side, with a search for behavioural models and styles that are "opposite" to those of the family.

The discovery, valorization, and adoption of environmental cues enable organizing one's life based on what works in one's relationship with other people, on common interests, and on the values that can be shared. A perception of the characteristics of the environment permits focusing on the parts of one's inner world that can be allowed to emerge, thus forming more or less close relationships.

Adolescent reorganization, which is characterized by the maturation of abstract skills and self-reflective thought, confers new features on the utilization of what is provided by the relational and socio-cultural environment, and entails a symbolic adjustment of the subjects' psychic world. Also in this case, when the subject has a good mentalization ability, the maintenance of affective, family and friendly references and the discovery of new significant relationships can become more adequate, profound, and flexible thanks to introspection and the use of symbolic functions, which enable a more specific and selective reading of internal states. Speculative competences on the sense of what one does and is also mature.

The ability to respond to semantic stress, related to situational judgements or to being stimulated to observe fair duties and commitments, prompts setting of the internal homeostasis on the basis of external signals. The reading of one's ability to respond to the tasks posed by the environment thus becomes the tacit basis to regulate achievable goals and significant relationships. The subject thus develops specific adaptive competences to manage the situations and problems that need to be addressed to achieve self-realization, on the basis of reliable attitudes and guidelines and through modulation and adjustment of the competences he/she has acquired over time.

The sense of wellbeing, corresponding with a state of psychic equilibrium, is achieved through the ability to modulate internal subjective states based on external information, which enables achievement of an increasingly mature tacit and explicit sense of the self and the world. This occurs in SRPO development pathways, both those where the care-giving environment appears "*highly available*" and "*accessible*" (as in *Contextualized* SRPOs) and those where it appears "*scarcely available*" and "*poorly accessible*" (as in *Normative* SRPOs).

B1. The Online Experience of Contextualized SRPOs

In a SRPO environment - where the care-giver appears close and available, provides reciprocity and communication ("*highly available*" and "*accessible*" care-giving), both over time, situation after situation, each time one feels its need ("*on-line*" care-giving), as well as in space, promptly offering advice and assistance ("*tight rein*" care-giving) - there develops an ability to recognize signs of approval, to select advantageous behaviors, and to predict the results that can be obtained and the merit judgments that can derive from them.

High, tight rein, on-line availability enables turning, when this is helpful, to the available reference persons for advice, to solve doubts, get confirmations (even in the form of constructive criticism), select goals that are suitable to one's attitudes, and form relationships inspired to mutual support, where one can exchange experience and assistance.

The child, through sensory-perceptual and emotional activations that match the attitudes, demands, and indications of significant figures, is guided to understand some aspects of the self and the world, learning to manage the former and to find a useful relationship with the latter. From the sum of these situational experiences the subject begins to identify the basic themes that need to be addressed to achieve a successful life through the indications, suggestions, comparisons, and tools that the significant figures provide each time. The very fact of making reference to the primary attachment figures to read the attitudes and expectations of the external world (thus updating the inner world and selecting the behaviour to be held) leads, since school age, to the identification of new significant references – peers as well as adults – as models and terms of comparison.

The stressor constituted by external *judgement* (i.e., internal representations of what the other may think and do, which are useful to gain an understanding of oneself) leads to develop as a key competence the ability to manage *personal exposure* to obtain the desired results. Social skills allow identification, and often also prevention, of *judgement* stress, and turning towards options that are within one's reach in the areas where one can be appreciated and valorized, without perceiving negative contexts as automatically reflecting inability or failure. In fact, in this adaptive route, positive and negative life events can be viewed as indicators of that which works or does not work, also in terms of possibility or impossibility of sharing situations and experiences; but this does not entail drawing automatically from the situation a putative personal value, as if the latter were a stock exchange share subject to market fluctuations.

The reading of experience by Contextualized SRPOs thus takes place through the development of three key *adaptive skills*: a) establishment of the goals that can be achieved and, consequently, commitment to meet them; b) identification (when needed) of those who will appreciate one, targeting them; and c) valorization of one's constructive and competitive abilities, bearing in mind external attitudes and judgements. As a result, management of experience by Contextualized SRPOs tends to involve development of two *abilities* that confer considerable advantages in terms of personal and social adaptation: the ability to "*focus on expectations*" by learning what one can or cannot expect from others, and a "*focus on results*", which reflects the subject's ability to choose the areas of experience that offer greater chances of success.

The resulting reading of experience is centred – in its immediacy of perceptions, images and subjective feelings – on a decoding in terms of confirmation/disconfirmation; in this SRPO the self-image in physical, aesthetic, professional, and relational terms and the adequacy of one's choices and objectives are processed in a subjective way and updated based on external judgements or comparisons with others.

In the *development routes* of Contextualized SRPOs, balanced care-giving/attachment enables achievement of good emotional and cognitive identification with parent figures and at the same time differentiation from them, and involves a similar behaviour also in subsequent relationships outside the family (with teachers, peers, friends, partners, colleagues). The child thus learns to decode the nuances of experience without drawing from this personal ambiguity or inconsistency, which are found in clinical disorders.

During *adolescent maturation*, interiorization of the confirmations the subject has received allows building a good sense of self and investing in the areas where he/she feels capable or talented. In addition, if the attachment processes have not been excessively invasive (in terms of anticipation and redefinition of his/her needs), the subject can learn to distinguish him/herself from others, and make the internal world emerge. This makes it possible to set future targets perceived as being within one's reach and to invest on figures that at the level of immediate experience appear to provide confirmations.

Therefore, in physiological development routes, the search for confirmation is an adaptive mode enabling the subject to identify and actively select the goals on which to invest and from which in turn to receive further confirmation. In such cases, any disconfirmation can be read and self-referred as a saddening negative event, whose effect on the internal equilibrium is however short-lived. Such disconfirmation can be overcome by honing more effective strategies to achieve one's goals or through the search for new goals that are more in line with one's way of being.

Significant *relational investments* are tacitly guided by the search for and achievement of appreciation, amiability and personal importance, as garnered from signals received from the figures perceived as significant. The couple alliance is put to the test and tacitly built based on the verification that the other is available and capable of providing confirmations that are important to the subject and on which the subject invests, and that he/she is capable gradually to accept or at least tolerate the weaknesses and deficiencies that he/she uncovers as intimateness increases and the subject's defences are lowered and increasingly disclose the way he/she really is.

With reference to clinical practice, in *Contextualized SRPOs* the reformulation of experience based on the significant episodes referred by the subject and analyzed in slow motion during psychotherapy, allows distinguishing the habitual way (experienced as the objective and only possible way) to self-refer experience from other possible ways that are consistent with the subject's PMO and are objectively more flexible and advantageous. This also allows harnessing the resources of the subject's PO, hitherto ignored or not used, for instance, the awareness of having a high sensitivity in focusing on attitudes and internal states of others allows using external judgements and acquired results as a compass to improve one's resources and focus constructively on the goals that are within one's reach, including the ability to take the others "the right way" without looking at them as into a mirror, or drawing from their attitude information on one's adequacy, amiability or value. At the same time, a better expression of one's resources allows bonding with friends, sharing affinities and significant aspects of one's world with those who can appreciate them, and constructing a couple relationship on a footing of adequate reciprocity, intimacy and exclusiveness.

B2. The Offline Experience of Normative SRPOs

On the other SRPO side - in an environment where the care-giver appears detached and poorly available with respect to one's inner needs, with scanty and measured reciprocity, fostering a responsibilization and maturation that highlight the respect of values and rules identified as important ("*scarcely available*" and "*poorly accessible*" care-giving), both over time, through interventions made only when they are felt to be required, irrespective of demands, and the setting of behavioral tasks that must always be carried out ("*off-line*" care-giving), and in space, through remote vigilance, checking that the child conforms without the need for point-by-point interventions ("*loose rein*" care-giving) - there develops an ability to become responsible by learning to respect rules and values, to do what is *right*.

Poor, loose rein, off-line availability allows the gradual discovery of one's resources and their use to support personal commitment, interiorizing and applying the lessons learned. Once identified, guidelines also serve to look for subjects they can be shared with, forming friendships based on shared feelings. The child discovers a kind of "main path" to follow, overcoming in this way limits and contradictions of life. To learn to recognize these aspects allows to focus on the problems of life, choosing the aspect that appears to be the best, initially based on the caregiver's indications and subsequently in an increasingly responsible and autonomous way. The possibility of making reference to specific interiorized criteria allows weighing the pros and cons of situations and to choose, when this seems necessary, among the contrasting aspects of reality that appear to be equivalent, without experiencing ambivalence (which occurs only in clinical disturbances).

The stressor constituted by *duty* (in the sense of doing what is right) leads to the development of the key skill of commitment as the ethical basis of one's behaviour. Based on the caregiver's indications, as the subject matures, he/she takes an increasingly active role in verifying, integrating or changing that which he/she should or should not follow. Whereas in Contextualized SRPOs external judgements, comparisons with others, and results achieved are used to test one's resources and competences in the case at hand, Normative SRPOs principally tend to draw from judgements and comparisons an affinity and the sharing with others of what the subject thinks counts in life. Personal commitment thus acquires a predominant role – as an expression of taking on personal responsibility – over results, which in turn do not depend exclusively on the subject, but also on external variables, and therefore cannot be fully ascribed to oneself.

The discovery of one's ability to conduct a responsible search for values and the behavioural rules connected with them allows discriminating what appears useful from what may be superfluous, overcoming the nuances of experience by making responsible choices (and avoid remaining trapped in them, losing oneself in an obsessively vacuous perfectionism). This tacit orientation also allows to search in the different experiences and seasons of life what may be valorized and what may be shared with those who express affinities with oneself.

In Normative SRPOs the reading of experience thus takes place via development of three key *adaptive skills*: a) self-commitment following the guidelines identified as a compass of one's conduct; b) identification (when needed) of those with whom motivations, values, ideals, and life choices may be shared; c) pursuance of the projects in which one believes without giving too much importance to comparisons or competition with others. As a result, to manage experience Normative SRPOs tend to develop two very useful personal and social adaptive *abilities*: "*a commitment founded on values*", which provide the guidelines to self-refer experience, and "*ethical solidarity*", based on the possibility of sharing right choices and projects on which it is worth investing.

The resulting reading of experience is centred – in its immediacy of perceptions, images, and subjective feelings – on the decoding of what one faces in terms of sufficient certainty, fairness, and scope for improvement, on the basis of which the subject draws the meaning of his/her choices and orientations, either personal (related to affection, job, speculation, and ethics) and relational.

In the *routes of development* of Normative SRPOs, the possibility of ordering and explaining contrasting aspects in a satisfactory way allows stabilization of the emotional setting and construction of the subject's identity in terms of acceptability and worth; experience is assessed explicitly in relation to how much it appears reliable and certain, based on the needs and values of which the subject feels a bearer.

If in infancy meeting the demands and criteria required for approval is the only possible way to gain access to the parents' care and affection, which are however always measured and controlled, during *adolescent maturation* the set of regulations and values that have been acquired are interiorized as personal active strategies, which stabilize and provide certainty in the face of the new scenarios opening for the subject. These strategies enable identification of the life choices that are in line with the subject's principles and of the affective investments from which he/she meets his/her need for certainty. However, viewing the route to experience as a difficult path – where commitment is crucial but results are not guaranteed – allows overcoming crises, doubts, and insecurity and to aim at achieving, to the extent that this is possible, realization at the professional and affective level to meet perceived needs. This knowledge strategy can thus find a keystone in discovering oneself as an active protagonist of one's experience and in planning projects and tasks considered important. This reinforces the sense of self and reliability, which could be undermined by the contradictory nature and the irreconcilable nuances perceived in reality.

Significant *relational investments* are tacitly guided by the attempt to share a life project according to the values perceived as right and unrenounceable. The couple alliance is tacitly founded on the identification of a figure that is available for and capable of sharing the values of union that the subject experiences as important, often after a fairly long period while the subject tests him/herself and the partner, to be certain that both are capable of taking on the responsibility entailed by loving one another and sharing a project.

The reformulation of experience based on the significant episodes referred by the subject and analyzed in slow motion during psychotherapy, allows distinguishing the habitual way (experienced as the objective and only possible way) to self-refer experience from other possible ways that are consistent with the subject's PO and are objectively more flexible and advantageous. This also allows harnessing the resources of the subject's PO, hitherto ignored or not used, for instance, the search for values and conducts that are consistent with them, which guide in making experience, allow overcoming life's problems and contradictions, and choosing travelling companions with whom it is possible to share what one believes in. This also regards – albeit on a level involving a greater commitment, responsibili-zation, and sharing – the more significant relationships, especially the affective ones, bearing in mind that for subjects with Normative SRPOs (unlike those with Controlling PRPOs) family ("*blood ties*") counts less than ethical and cultural similarity ("*ideal ties*").

3. Psychotherapeutic Implications in the Reformulation of Experience

The assimilation and reprocessing of experience according to the route of development of individual adaptive competences allows focusing on the physiological resources provided by each subject's PO. In fact, each PO allows developing some *specific sensory-perceptual susceptibilities*, based on the fact that it appears more "*permeable*" and "*sensitive*" to the more significant aspects and to the *predominant domains* in its environment. In this way each PO evolves some distinctive assessment and operational strategies, and achieves selective and specialized adaptive resources that are then honed in the course of life.

These physiological adaptive resources continue somehow to be present (and thus need to be focused) in the different forms of clinically significant conditions. In fact, in pathological conditions these resources appear to be masked by psychic and behavioural disturbances, which confine and limit the subject and undermine the search for a better adaptation, preventing the exploitation of the potential they nonetheless have. In this respect, psychopathology induces a variably profound restriction of the modes of self-referring experience, which the subject lives as objective, external, and thus suffers passively, with a consequent damage to personal adaptation resources. The psychotherapeutic process, through the reformulation of experience, has the potential to focus on and manage these modes of self-referring experience, changing the relevant emotional activations, and enabling harnessing of the subject's resources in a more flexible way and one that is more adaptively advantageous.

The possibility of taking into account the tacit and explicit resources characterizing the subject's PO enables tailoring the therapeutic process and places the subject at the centre of the route, for which he/she provides the internal material, so that he/she can begin to explore it with the tools provided by the therapist. A key aspect of the reformulation of experience it is thus the evaluation of the subjective modes of assimilating experience, that have consolidated to define the subject's sense of self. The discovery of new modes of assimilating and self-referring experience – through the reformulation of life episodes analyzed in slow motion in the psychotherapy sessions – is essential to find new modes that are consistent with one's PO and are more adaptive and flexible. This makes it possible for the subject to achieve a greater separation between his/her internal world and that of others, and change the emotional activations stemming from the same experience, which was earlier assimilated in the only way that appeared possible as if it were the sole objective way.

Reformulating experience thus makes available to the subject new instruments to self-refer what has happened, producing a different emotional activation that the subject – now aware of being the protagonist of his/her internal world – can experience as his/her own, discovering his/her ability to valorize his/her own *uniqueness* also when facing negative experiences, without necessarily drawing from them a sense of personal inadequacy. The modalities of one's PO, earlier experienced as something that involved suffering the events passively, begin to be guided by the internal world of the subject, who no longer identifies history (experience) with personal worth (in terms of abilities and amiability). The analysis in slow motion of one's way of functioning therefore allows breaking the correspondence equating events that have happened and personal worth. This leads to the discovery of new ways to express one's uniqueness, of which the subject realizes he/she is the creator, since he/she no longer passively suffers experience. In fact, that negativity that had hitherto been experienced as an objective fact begins to appear as an actively contrived construction, originated from the habitual mode of self-referring experience whereby a negative situation automatically generated a personal negate effect.

The discovery of *new ways of self-referring experience* thus opens new prospects for testing one's ability to function and be in the world, also extending and changing the subjective palette provided by emotional activations. The latter are clearly different if the same experience is lived no longer as further proof of one's putative negative value, but simply as one of the many things that may go wrong in a person's life (and which at the tacit level can actively be produced to test that single mode of activation in which the subject has learned to recognize him/herself). This makes available the adaptive tools provided by the subject's PO, which begins to prove a positive resource that enables focusing on and managing specific aspects of internal and external reality, identifying the behaviours and choices that are more useful for self-realization in the current stage of life.

In conclusion, the reformulation of experience can help identify and bring forth the potential resources and adaptive competences – inhibited at the tacit level and ignored at the explicit level – that are related to the PO of the subject undergoing therapy. It is possible to see how on the one hand there is a constancy related to the consistency of development processes that lead to a specific basic organizational closure (Controlling, Detached, Contextualized or Normative types); on the other hand, it is possible to observe the complex evolutionary dynamics connected with the adaptive abilities used to assimilate and self-refer experience. The focus through contrast adjustment, achieved by comparing the sensory-perceptual emotional flow and the ideo-affective schemes used to sustain internal consistency and personal identity, prompts the emergence of the subjective modalities by which each subject articulates his/her sense of self in a unique, unitary and global way. The reformulation of experience thus provides access to new ways of living it that are consistent with the subject's PO, thus also changing the matching behavioural expressions and highlighting those subjective resources that make each individual unique.

References

Guidano V.F.: *Complexity of the Self*. Guilford, New York, 1987.

Guidano V.F.: *The Self in Process*. Guilford, New York, 1991.

Guidano V.F.: *Trascrizione del Training di Psicoterapia Cognitiva Post-Razionalista.* Roma, 1999 [*Lessons from the Post-Rationalist Cognitive Training. Rome,* 1999].

Guidano V.F.: *Le Dimensioni del Sé. Una Lezione sugli Ultimi Sviluppi del Modello Post-Razionalista* [*The Self Dimensions. A Lesson on the Last Development of the Post-Rationalist Approach*]. Alpes, Roma, 2010.

Nardi B.: *CostruirSi. Sviluppo e Adattamento del Sé nella Normalità e nella Patologia* [*Self Constructing. Development and Ability to Adapt of the Self in Physiological and in Psychopathological Conditions*]. Franco Angeli, Milano, 2007.

Nardi B.: *Guidelines on the construction of a post-rationalist therapeutic approach.* European Psychotherapy 9 (1), 57-67, 2010.

Nardi B.: *La Coscienza di Sé. Origine del Significato Personale* [*The Self Consciousness. Origin of the Personal Meaning*]. Franco Angeli, Milano, 2013.

Nardi B.: La conoscenza tacita nelle organizzazioni inward e outward: lo sguardo oltre la siepe. In: Nardi B., Arimatea E. (Eds.), *Lavorare con la Conoscenza Tacita. Atti del XIV Convegno di Psicologia e Psicopatologia Post-Razionalista*, pp. 14-32 [*Working with the tacit experience. Proceedings of the XIV Meeting of Post-Rationalist Psychology and Psychopathology*, pp. 14-32]. Accademia dei Cognitivi della Marca, Ancona, 2013.

Nardi B.: Adaptive specificity of the self and complexity of the inward and outward Personal Meaning Organizations (PMO), special issue on Psychotherapy as complex dynamical system (Orsucci F. Ed.). *Chaos and Complexity Letters,* 8 (2-3), 2015.

Nardi B.: *Esperienza Soggettiva e Organizzazioni di Personalità* [*Subjective Experience and Personality Organizations*]. Accademia dei Cognitivi della Marca, Lopez, Ancona, 2016.

Nardi B., Arimatea E., Giovagnoli S., Blasi S., Bellantuono C., Rezzonico G.: *The Mini Questionnaire of Personal Organization (MQPO): preliminary validation of a new Post-Rationalist Personality questionnaire.* Clinical Psychology and Psychotherapy, DOI 10.1002/CPP.740, 19, 78-90, 2012.

Nardi B., Capecci I., Fabri M., Polonara G., Salvolini U., Bellantuono C., Moltedo A.: Estudio mediante imagen funcional de resonancia magnética (fMRI) de las activaciones emotivas correlacionadas a la presentación de rostros extraños o del propio rostro en sujetos con personalidad inward y outward [fMRI investigation of emotional activations during visual processing of others or own facial expressions in subjects with an inward or outward personality]. *Revista Chilena de Neuropsiquiatría*, 46(3), 168-181, 2008.

Nardi B., Marini A., Turchi C., Arimatea E., Tagliabracci A., Bellantuono C.: *Role of 5-HTTLPR polymorphism in the development of the inward/outward personality organization: a genetic association study.* Plos One, 8 (12), e82192 1-8, 2013.

Nardi B., Moltedo A.: *Rol de la relacion de reciprocidad en el desarrollo de las diversas organizaciones de significado personal* [*Rule of reciprocity in the development of the Personal Meaning Organizations*]. Gaceta de Psiquiatria Universitaria (Universidad de Chile, Facultad de Medicina), 4(4), 3, 345-358, 2008.

Nardi B: *Organizzazioni di Personalità. Normalità e Patologia Psichica.* [*Personality Organizations: Normality and Mental Disorders*]. Itard, Chiaravalle (AN), 2020.

Technical Chapters

Chapter 4

The Chaos and Complexity of Athletes' Emotional Experiences: The Insynergy Sport Fast Emotional Insight (ISFEI) Game Cards*

Francesca Stecchi*
and Silvia Leone
InSynergy MindLab, Padua, Italy

Abstract

This work introduces the InSynergy Sport Emotional Insight (ISFEI Game Cards) as an explorative counselling framework with the aim of assisting sports psychology practitioners in understanding how athletes experience emotions. This tool is designed to quickly elicit clients' deep experience narratives in non-clinical settings. The authors first explain how the ISFEI Game Cards was conceived; they then introduce the Chaos and Complexity Theory, as the theoretical foundation of this framework, and describe both the structure and the recommended method for the use of ISFEI. Finally, clients' feedbacks and future trends are reported as starting points for improving the framework.

Keywords: emotions, chaos and complexity theory, play therapy, explorative counselling framework, sports, athletes

* This is a reformatted version of an article previously published in Chaos and Complexity Letters, Volume 16 Issue 3 (2022), New York: Nova Science Publishers.

* Corresponding Author's E-mail: francesca.stecchi@insynergy.it.

In: Complexity Science in Human Change
Editors: Franco F. Orsucci and Nicoletta Sala
ISBN: 979-8-89113-408-9

Introduction

The relationship between performance and emotional arousal has gained importance over time in the field of sports psychology (Gould et al., 2002). Academics have used different theoretical frameworks to explain how emotions can affect the behaviour during competitions and its outcomes. According to research, emotional states can predict athletic performance (Hanin, 2000; Robazza et al., 2006). In particular, the ability to adopt emotional regulation strategies (Gross and Thompson, 2007) determines physiological, behavioural, cognitive, and subjective experience responses. The sports psychology intervention programmes currently in use also aim to provide athletes with the best resources for improving their emotional control under pressure. Elite athletes are tested in a variety of circumstances, and frequently, the factors that affect their success can be found in areas outside than sports. With time, research has adopted a holistic perspective of the athlete, whose subjective experience is divided into multiple, interlocking planes. The sportsman is a person first, before being an athlete.

Adopting a holistic lifespan perspective implies not only considering the athlete as a "whole being;" it may allow to identify emotions, moods, and thoughts linked to a life experience, that may unconsciously affect the sports career. According to this viewpoint, the identification of emotional domains with negative and positive valence would contribute to the success of the intervention. Understanding the athlete's barriers and strengths seems to be the starting point of this analysis process. However, sports psychology practitioners are often faced with the impossibility of undertaking such an in-depth analysis of the athlete's emotional context (e.g., due to the need to conduct a quick, short-term intervention or the lack of a background in clinical psychology).

Therefore, the aim of this explorative counselling framework is to introduce the use of the "InSynergy Sport Fast Emotional Insight (ISFEI) Game Cards" as a tool to support sport psychology practitioners in gaining insight on the elite athletes' emotional experience at the onset of a non-clinical intervention. To do so, this work discusses (a) the context in which ISGC was conceived; (b) the theoretical foundations driving the development of the card decks; (c) the methodological guidance on how to implement ISGC within a non-clinical intervention; (d) the main findings from using ISFEI with a pilot sample of athletes; and (e) the limitations and ethical considerations to deploy ISFEI.

Why the InSynergy Sport Fast Emotional Insight (ISFEI) Game Cards Was Developed

In recent years, the necessity to prioritise treating athletes as people first and foremost has increased academic interest in applied sports psychology. The emphasis on the binomial "cognitive ability/performance" has really changed more and more in favour of "performance/well-being," where the athlete is only one aspect of the person who also consists of emotions, thoughts, and a private life. This all-encompassing perspective would allow the athletes to adhere to the deepest part of themselves and, consequently, to improve their performance. Numerous biopsychosocial factors, that can either favourably or negatively impact the individual development, filter the athlete's evolution and the attainment of a great performance.

In this regard, Bond (2001) legitimised the decision to consider the athlete first and foremost as a person; hence the urgency to promote the optimal functioning of the individual through specific intervention programmes. Soon, sports psychology consultants and practitioners would incorporate programmes aimed at monitoring elite athletes' lifestyles, group dynamics, and non-sporting domains into their psychological development interventions. The study of sport psychology has been encouraged by the recognition of sport as a "*multidimensional phenomenon*" (Friesen and Orlick, 2011), which operates within an ontological relativism framework (Smith and Sparkes, 2009). The athlete-client is thus given the responsibility of acting as an agent inside the own psychological reality; a subjective state dependent on the quality of thought (whereby quality, in purely psychological terms, we mean only the difference inside the inter-individual content). As a result, the urge to enhance the person-athlete's mental health and well-being grows alongside more structured interventions that embrace a holistic lifespan approach (Stambulova, 2010).

However, the lack of a same background in clinical psychology for all the sports psychology practitioners does not always allow for the application of functional methods and techniques to the emotional analysis of clients. This work is certainly not meant to be a venue for debate (which is still open in the field of sports psychology), but rather merely to highlight the importance of dealing with emotions, both in short-term and long-term interventions. In order to assist athletes in exploring their innermost dynamics and becoming more conscious of their complicated emotional experiences, this research is an explorative counselling framework that starts with the need to explore the secret to a quick and effective response.

Theoretical Foundations of ISFEI Game Cards

The difficulty encountered in the analysis of emotions and human agency is grounded in the principles of Morin's Complex Systems (2008). Chaos and Complexity Theory is in fact the main theoretical framework used in this study. A set of mathematical concepts defines, in this case, not only the complexity of existing systems, but also tries to explain how these change over time. In recent years, several authors have attempted to apply this theory in the field of psychology and, only more recently, in the field of sport (Carbinatto et al., 2011; Balagué et al., 2013). The applicability of non-linear systems theory in psychology is only possible by moving away from the reductionism that scientific research has been advocating in a field where there is still much to be discovered. The predictivity of apparently random events has always been debated in psychopathology and psychiatry; assuming the existence of complex, dynamic, and superimposable systems presupposes the search for specific mathematical functions, that contribute to the prediction of the variables in question (Haken, 1983; Schöner and Kelso, 1988). Although dynamic patterns (e.g., - in the psycho-pathological field - depression) are recognised as expressions of a single system, these are emergent distinct phenomena (Anderson, 1972) and self-organised (Haken, 2000). Therefore, a dysfunctional behaviour, emotion or cognitive pattern would become the attractive units of a central pole, a source that can be traced by mathematical models. Studies in the field of social cognitive psychology have applied the principles of complex systems to understand how best to structure coaching (Bowes and Jones, 2006). Just like the coach, the athlete would be part of a complex system, of which all possible variables that contribute to the performance outcome should be highlighted. The multidimensionality of sport forces the athletes and those around them into complex dynamics and relationships (all of which are reflected in the person's psyche). Sporting performance is in fact the behavioural outcome of the athletes, who become the main actors in their life. By this we mean emphasising the active role of the person, whose body becomes biologically and psychologically a reflexive and active being. This vision is suited to a type of counselling intervention, which favours the active mental and emotional participation of the athletes, who - we think - need to become fully aware of their internal and external dynamics.

The difficulty of predicting behaviour and emotions in sport is also and above all reflected in the need to structure short-term interventions. Elite athletes are in fact constantly busy preparing for competitions; very often sports psychology practitioners find themselves with little time to structure targeted and effective interventions before competitions.

This study is therefore best be seen as a need to provide sports psychologists with a quick and practical tool for identification and emotional analysis. From the experience of clinical consultations with athlete and non-athlete clients, the concept of reducing potential sources of stress during consultation was developed. In line with the theory of non-linear systems "*Synergizing, instead of training, defines an improved understanding of the process and helps scientists, coaches, and practitioners to create safer and effective interventions*" (Pol et al., 2020, p. 1).

The Structure of the ISFEI Game Cards

Theoretical Structure and Contents of the ISFEI Game Cards

The "InSynergy Sport Fast Emotional Insight (ISFEI) Game Cards" stems from the authors' practice experiences in clinical and applied sport psychology. This framework is designed to help sports psychology practitioners access the athletes' emotional experiences more easily and quickly. The structuring of the ISFEI Game Cards took place in two phases, with a total duration of six months. Italian Elite athletes and new clients receiving psychological counselling interventions were the subjects that signed up for the programme. All the clients signed two informed consents: one guaranteeing their right to privacy and the possibility of being subjected to assessment tools (as established by Art. 1 of Law 56/89 of the Order of the Profession of Psychologist); a second informed consent to reaffirm the aims of the study and the possibility of using the clients' feedbacks for scientific research purposes. The study was conducted according to the guidelines of the Declaration of Helsinki.

In a first phase, the effectiveness of the "Dixit" playing cards (Jean-Louis Roubira, 2008) was tested for three extensions of 84 cards each "Revelations-2016" (illustrated by Jean Louis Roubira and Marina Coudray), "Dixit

Journey-2012" (illustrated by Xavier Collette), and "Dixit Memories" (illustrated by Carine Hinder and Jèrôme Péllissier), for a total of 252 illustrations. The training lasted 3 months.

Playing cards were administered to the clients so that they perceived them as a less invasive tool. Therefore, an approach that gave them the feeling of not being assessed or diagnosed was preferred; these meetings were thus transformed into "a game within a game." Indeed, the literature reports the effectiveness of the "Play Therapy" within the clinical setting: play would seem to facilitate and reinforce the power of therapy (Ray et al., 2001; Klein, 2010; Schaefer and Drewes, 2011).

Schaefer and Drewes (2011) identified twenty fundamental therapeutic powers of play. These include agents of change, self-expression, emotion regulation, resilience, self-esteem, and stress management. Furthermore, many studies emphasise the role of play in fostering the development of empathy (Brownell et al., 2002).

Clients were asked for feedbacks at the end of each session. The final interview followed a fixed pattern, from which relevant information about the effectiveness of the cards would be gathered. The client was asked how they felt at the end of the session and which elements in the illustrations had captured their attention the most. All the clients involved stated that, thanks to the use of the cards, they were able to verbalise emotions that they would hardly have expressed in that context and so quickly; they also stated that those same emotions represented a present experience. This last element was interpreted as a strength of the intervention. The use of the illustrations could in fact have evoked negative emotions linked to the clients' past; such a situation could also have required a more in-depth psychological assessment and, in particular cases, referral to a psychotherapist. Interestingly, during the sessions, all clients followed a similar behavioural pattern when presenting each card. A natural description of the card followed an initial verbal closing approach, surrounded by moments of reflection and silence. Suddenly a shift occurred; during the description, the object of the card became a self-representation. Only at this stage, the clients would move into the re-enactment of a situation triggering the present emotional experience.

The unexpected emotional richness of the clients' stories allowed the authors to structure a second phase of intervention.

Three more months were dedicated to the administration of a set of 15 illustrations (see Appendix 1), kindly provided by a professional illustrator. These pictures are a test to see if their style, colours, and depictions evoke deep experiences like the Dixit Cards previously tested. However, these illustrations are only the first step towards the final version of this counselling framework. In order to obtain a real structure of the ISFEI Game Cards, the authors envisage the creation of two further sets of images to be tested (15 cards in each deck, for a total of 30 cards); the outcome on clients of the use of these illustrations will be compared with the 15 already available images.

The comparison will be based on three different characteristics: illustrative style, colours, and content (the object).

The first pack of illustrations contained illustrations selected from the illustrator's existing works; these were therefore not created on the basis of specific rules.

The second will use different representations with special attention to colours, to observe whether the client will be more attracted to the figure or the colour. For this purpose, Luscher's Colour Theory (1969) will be used to assign colour dominance to each card; combinations will be created between the "emotional meaning of the colour" and the "object" of each card.

The third group of cards, on the other hand, will contain some specific content referring to selected emotional and psychological domains. The most common basic emotions and psychological constructs reported in the sport psychology literature were chosen. In particular, much space was given to pre-competitive anxiety (Jones, 1995), anger (Davis et al., 2010), fear (Zhang et al., 2018), frustration (Haraldsen et al., 2020), relationship with the coach (Nicholas et al., 2019), self-confidence (Feltz, 2007), domains of self (self-image, ideal-self, and self-esteem) (Feltz, 2007; Brunet et al., 2012; Ouyang, 2020), self-efficacy (Sivrikaya, 2019), and coping strategies (Nicholls et al., 2007). None of the 45 cards contain objects related to the sportive world; further cards will be created with specific references to sport, in order to test a possible difference in the evocative power between representations of experiences closely related to sport and more dreamlike images of different nature. The first deck was the only one to have been administered in the last three months of the intervention. All other cards will be presented to the clients in the future, in order to understand which details of the cards have greater evocative power. Indeed, the authors wish to point out that this is an explorative counselling framework and, as such, needs further trials to better understand which card structure is most appropriate to use in the future.

Tested and Suggested Methodology

The cards were always administered no earlier than the second session. The reasons for this choice can be traced back to the authors' desire to put the client at ease; to allow the practitioner to understand the client's needs; to better understand the client's experiences in order to be able to select a limited number of cards, in line with the narrative emerged during the first counselling session. The structuring of the administration method followed several attempts and observations during the sessions. Hence, the pattern that the authors recommend using will be presented. The effectiveness of the proposed scheme stems from the clients' feedbacks in the individual interviews.

Below, the steps for the use of the ISFEI Game Cards:

1. The practitioner selects 25 relevant cards, based on the client's tale. The client is asked to choose 5 cards from those proposed and to place them in order of importance. The client's story can now begin, starting with the card in the first position and continuing with the remaining 4 cards.
2. The administration of the cards is interrupted for at least a week, to avoid the habituation effect to the cards and so that the client can rework the emotional states exhibited in both the sporting and private spheres of life.
3. The second administration of the deck takes place differently. The psychologist selects 5 cards based on his/her clinical experience and the client's previous tale. The practitioner lays the covered cards on the table and lets the client choose one of the 2 following methods:
   - Take one card at a time, describing the mood and emotions aroused.
   - Take all five cards and look at the illustrations; the client then can decide whether to change a card with a covered card from the whole deck. The card that the client chooses to change will be offered later during the same session or represented in subsequent sessions. The sessions showed that clients acted out an avoidance of an emotional experience evoked by the replaced card. Dysfunctional coping strategies, such as problem avoidance, could negatively influence the athlete's mental health and contribute to poor sports performance (Hanin, 2010).
4. Should the interview with the client reach a "dead end" and a lack of dialogue, the practitioner can administer the cards in the following

way, in order to make the most of the time available: practitioners divide the available deck into two parts; and then give the first half of the cards to the client. They randomly remove a card from the deck and display it to the client with the following request: "Choose now, from your deck, the card that you think matches the one I removed." In this way the client acts by free association (Koukourikos et al., 2020) and is automatically led to expose an emotional experience in a less stressful way.

Clients' Feedbacks

During the six months of training, the clients revealed that they benefited enormously from using the cards. They confirmed the usefulness of the Dixit cards, and later the ISFE Game Cards, by reporting feelings and emotions experienced while using this framework. Through the cards, individuals were able to free themselves more easily from intrusive thoughts and intense emotions; in other circumstances, they would have felt shame and fear in expressing their internal experiences. Moreover, they report having brought back such profound and almost "forgotten" personal events. The visual and card-playing components seem to be the strong point of this framework; without visual stimuli and game modes, many clients would not have expressed inner contents that are fundamental to the success of the intervention.

In particular, the illustrations of the ISFE Game Cards tested in the second period gave satisfactory results.

The authors leave open the possibility that these cards, despite their effectiveness, express totally different emotional experiences for each subject. To date, clients who participated in this study have reported different emotions for the same cards presented and evoked life events of different intensity and emotional value. Furthermore, a same object could be interpreted differently depending on individual life experiences; an illustration was perceived by a tennis player as a "*trophy*» and by a gymnast as *«a woman's body... Something I feel worried and uncomfortable about."* Many of the participants were also particularly impressed by the colours of the pictures, while others only by the object represented. Furthermore, experiences with clients in the use of ISFEI Game Cards revealed a key component: the cards elicited experiences almost always refer to a present dimension.

Finally, ISFEI Game Cards could be an easily applicable tool in interventions with sports teams. In contrast to working with individuals, sessions with teams may be more difficult to manage; the size of the group places the practitioner in front of a greater number of dynamics to deal with and variables to control (Cotteril, 2013). As is very often the case in interventions with teams, in the limited time of counselling sessions it is difficult to deal with every aspect in such depth; identifying the problem immediately is not always possible. Similarly, following the needs of the individual athletes, that make up the team, would not be so easy. The ISFEI Game Cards, on the other hand, are an excellent ally of time and short-term interventions; this method may allow the team members to be coordinated by subjecting them to a game (just what they are used to, a team game); communicating thoughts and expose problems related to group dynamics would be possible in this way.

Conclusion and Future Trends

This explorative counselling framework was created with the aim of providing sports psychology professionals with a tool that is simple and fun to use with athlete-clients.

Experiences with clients in the use of ISFEI Game Cards have brought to light a fundamental element: the experiences elicited by the presentation of the cards are almost always referable to a present dimension. This allows to quickly investigate the athlete's cognitive and emotional states in both pre-competition and post-competition settings. The accuracy and effectiveness of this tool, which will be further tested and improved in the coming months, derived from the athletes' feedbacks and the positive outcomes obtained in clinical practice work with clients. The authors were able to ascertain that the ISFEI Game Cards, presented in the form of a game, encourage clients to express moods and experiences related to the present in a simple and immediate way. In fact, the majority of clients stated "*I would never have dealt with this topic without the use of the cards*"; the reasons for this can be traced back to several factors easily found in counselling. Individuals tend to focus on a specific story within the hour of the session. This tale is generally the most "urgent" of the day. ISFEI Game Cards come to the practitioner's rescue, promoting the recollection of emotional experiences that clients do not remember or do not consider important enough. However, there are limits to the use of ISFEI Game Cards. The first limitation is the clients' handling of intense emotions; the authors recommend, for purely ethical reasons, the use of the cards to professionals with a background in clinical psychology. Many of the clients may verbally report traumatic experiences, such as physical and sexual abuse; in cases such as these, the authors recommend that professionals without a psychotherapeutic background refer them to specialist colleagues who can deal with the case.

Finally, despite the results obtained, this explorative counselling framework is still a work in progress; the ISFEI Game Cards is a tool to be completed and tested in the future. The authors intend to add many more cards based on future feedbacks. In this way, the content of the cards in terms of colours and content will be more effective, in line with the sports context and the athletes' needs.

Author Contributions

F. S. has conceived and tested this counselling framework on clients within her clinical practice. F. S. self-funded this study. F. S. and S. L. contributed to the writing of the explorative counselling framework. All the authors drafted this manuscript, critically revised it for important intellectual content, read it, and approved the final manuscript.

Acknowledgments

We would like to thank Ann Nadine (https://www.ann-nadine.com/)for providing the project with her artistic know-how through the illustration of the ISFEI game cards. We would also like to thank all the clients who, by their feedbacks and enthusiastic participation, have contributed to the realization of this project. The authors declare they have no conflict of interest with this statement. The study was self-funded. Informed consent was obtained from all subjects involved in the study. The study was conducted according to the guidelines of the Declaration of Helsinki.

Appendix 1

References

Anderson, P. W. (1972) More Is Different. *Science*, 177, 393-396.

Balague, N., Torrents, C., Hristovski, R. Keith Davids and Duarte Araújo (2013). Overview of complex systems in sport. *J. Syst. Sci. Complex* 26, 4–13.

Bond, J. (2001). The provision of sport psychology services during competition tours. In G. Tenenbaum (Ed.), *The practice of sport psychology* (pp. 217-229). Morgantown, WV: Fitness Information Technology, Inc.

Bond, J. (2002). Applied sport psychology: Philosophy, reflections, and experience. *International Journal of Sport Psychology*, 33, 19-37.

Bowes, I., & Jones, R. L. (2006). *The Sport P*sychologist, 20, 235-245 © 2006 Human Kinetics, Inc.

Bowes, I., & Jones, R. L. (2006). Working at the edge of chaos: Understanding coaching as a complex, interpersonal system. *The Sport Psychologist*, 20(2), 235.

Brownell, C. A., Zerwas, S., & Balaram, G. (2002). Peers, cooperative play, and the development of empathy in children. *Behavioral and Brain Sciences,* 25, 28–29.

Brunet, J., Sabiston, C. M., Castonguay, A. L., Ferguson, L., & Bessette, N. (2012). The association between body-related self-discrepancies and women's physical activity: The mediating role of motivation. *Journal of Sport & Exercise Psychology*, 34, 102–123.

Carbinatto, M. V., Moreira, W. W., Nunomura, M., Tsukamoto, M. H. C., & Nista-Piccolo, V. L. (2011). Paradigms shift in sport sciences: Body's focus. *World Academy of Science, Engineering and Technology*, 80(8), pp. 1090–1094.

Cotterill, S. T. (2013). *Team psychology in sports: Theory and practice*. Hove, EastSussex: Routledge.

Davis, P. A., Woodman, T., Callow, N. (2010). Better out than in: The influence of anger regulation on physical performance. *Personality and Individual Differences*, 49(5), 457-460.

Feltz, D. L. (2007). Self-confidence and sports performance. In D. Smith & M. Bar-Eli (Eds.), *Essential readings in sport and exercise psychology* (pp. 278–294). Human Kinetics.

Friesen, A., & Orlick, T. (2010). A qualitative analysis of holistic sport psychology consultants' professional philosophies. *The Sport Psychologist*, 24, 227-244.

Gould, D., Greenleaf, C., & Krane, V. (2002). Arousal-anxiety and sport behavior. In T. S. Horn (Ed.), *Advances in sport psychology* (pp. 207–236). Human Kinetics.

Gross, J. J., & Thompson, R. A. (2007). Emotion Regulation: Conceptual Foundations. In J. J. Gross (Ed.), *Handbook of emotion regulation* (pp. 3–24). The Guilford Press.

Haken, H. (1983). *Synergetics: An Introduction*, E & Springer, Berlin.

Haken, H. (2020) *Information and Self-Organization. A Macroscopic Approach to Complex Systems*, E & Springer, New York.

Hanin, Y. (2010). "Coping with anxiety in sport," in *Coping in Sport: Theory, Methods, and Related Constructs.* Ed. A. R. Nicholls (Hauppauge, NY: Nova SciencePublishers inc), 159–175.

Hanin, Y. L. (2000). Successful and poor performance and emotions. In Y. L. Hanin (Ed.), *Emotions in sport* (pp. 157–187). Human Kinetics.

Haraldsen, H. M., Solstad, B. E., Ivarsson, A., Halvari, H., & Abrahamsen, F. E. (2020). Change in basic need frustration in relation to perfectionism, anxiety, and performance in elite junior performers. *Scand. J. Med. Sci. Sports*, 30(4), 754-765.

Jones, G. (1995). More than just a game: Research developments and issues in competitive anxiety in sport. *British Journal of Psychology*, 86(4), 449–478.

Klein, M. (2010) The psychoanalytic play technique. *American Journal of Orthopsychiatry* 25(2), 223-237.

Koukourikos, K., Tsaloglidou, A., Tzeha, L., Iliadis, C., Frantzana, A., Katsimbeli, A., & Kourkouta, L. (2021). An Overview of Play Therapy. *Mater. Sociomed.*, 33(4):293-297.

Luscher, M., & Scott, I. A. (1969). *The luscher color test.* Pocket Books.

Morin, E. (2008). *On complexity.* Cresskill, NJ: Hampton Press.

Nicholls, A. R., & Polman, R. C. J. (2007) Coping in sport: A systematic review. *Journal of Sports Sciences*, 25(1), 11-31.

Nicolas, M., Gaudreau, P., & Franche, V. (2011). Perception of coaching behaviors, coping, and achievement in a sport competition. *J. Sport Exerc. Psychol.*, 33(3), 460-8.

Ouyang, Y., Wang, K., Zhang, T., Peng, L., Song, G., & Luo, J. (2020). The Influence of Sports Participation on Body Image, Self-Efficacy, and Self-Esteem in College Students. *Front. Psychol.*, 10, 3039.

Pol, R., Balagué, N., Ric, A., Torrents, C., Kiely, J., & Hristovski, R. (2020). Training or Synergizing? Complex Systems Principles Change the Understanding of Sport Processes. *Sports Med. Open*, 6(1):28.

Ray, D., Bratton, S., Rhine, T., & Jones, L. (2001). The effectiveness of play therapy: Responding to the critics. *International Journal of Play Therapy*, 10(1), 85–108.

Robazza, C. (2006). Emotion in sport: An IZOF perspective. InS. Hanton & S. D. Mellalieu (Eds.), *Literature reviews in sport psychology* (pp. 127–158). New York: Nova Science.

Schaefer, C. E., & Drewes, A. A. (2011). The therapeutic powers of play and play therapy. In C. E. Schaefer (Ed.), *Foundations of play therapy* (pp. 15–25). John Wiley & Sons Inc.

Schöner, G., & Kelso, J. A. (1988). Dynamic pattern generation in behavioral and neural systems. *Science,* 239(4847), 1513-20.

Sivrikaya, H. (2019). The Role of Self-efficacy on Performance of Sports Skills of Football Players. *Journal of Education and Training Studies*, 6(12), 79-75.

Smith, B., & Sparkes, A. C. (2009). Narrative inquiry in sport and exercise psychology: What can it mean, and why might we do it? *Psychology of Sport and Exercise*, 10(1), 1–11.

Stambulova, N. (2010). Counseling athletes in career transitions: The five-step career planning strategy. *Journal of Sport Psychology in Action*, 1(2), 95–105.

Zhang, S., Woodman, T., & Roberts, R. (2018). Anxiety and Fear in Sport and Performance. In *Oxford Research Encyclopedia of Psychology* Oxford University Press.

Chapter 5

An Experimental Verification of NeuroPhysics Treatment by Executing a Multifractal Analysis of Surface Electromyography Signals on Trapezius, Abdominals, and Adductor Muscles in Athletes

Calogero Rinzivillo[1]
Ferda Kaleagasioglu[2,3,*]
Francesco Casciaro[4]
Fabio Scoppa[5]
Riccardo Marvulli[6]
Ken Ware[7]
Elio Conte[3]
and Antonio Di Ieva[8]
[1]Department of Medical-Surgical Sciences and Advanced Technologies, Medical School and Policlinic, University of Catania, Italy
[2]Department of Pharmacology and Clinical Pharmacology, Faculty of Medicine, Istinye University, Istanbul, Turkey
[3]School of Advanced International Studies on Applied Theoretical and Non-Linear Methodologies of Physics, Bari, Italy
[4]Department of Oncology and Internal Medicine, University Aldo Moro, Bari, Italy
[5]Faculty of Medicine and Dental Surgery, Sapienza University of Rome, Rome, Italy
[6]Department of Basic Science, Neuroscience and Sense Organs, University Aldo Moro, Bari, Italy
[7]Institute Neurotricional Sciences Pty. Ltd., Gold Coast, QLD, Australia

* Corresponding Author's Email: ferda.kaleagasioglu@istinye.edu.tr.

In: Complexity Science in Human Change
Editors: Franco F. Orsucci and Nicoletta Sala
ISBN: 979-8-89113-408-9

[8]Computational NeuroSurgery (CNS) Lab, Macquarie Medical School, Faculty of Medicine, Health and Human Sciences, Macquarie University, Sydney, Australia

Abstract

Background: NeuroPhysics Treatment (NPT) is an exercise-based approach that was hypothesized to utilize the body's genius to trigger and organize self-healing. NPT involves mild resistance training exercises with extremely slow movements in an erect posture under therapeutic coaching. NPT was hypothetically based on system dynamics, chaos, and fractal theory. The aim of this study was to investigate this thesis by a cross-disciplinary approach.

Methods: The differences between surface electromyography (sEMG) signals recorded on the trapezius, the abdominals, and the adductors of 10 healthy, male athletes at rest, during and following NPT were investigated by using the multifractal analysis based on Multifractal Detrended Analysis and with added Surrogate Data Analysis.

Results: All the muscles represent complex systems, and their bio signal profiles constitute a multifractal. During the NPT treatment, these systems are subjected to continuous and differential transitions, enabling a continuous and differential rearrangement of the indices between different muscles and in the same muscles.

Conclusions: This study is methodological and evidences the importance of always performing the multifractal analysis in sEMG studies and indicates the necessity for the same analysis of the EEG and of the R-R signals to obtain a whole picture of the involved dynamics, as a future prospect.

Keywords: surface electromyography, multifractal analysis, detrended fluctuation analysis, NeuroPhysics treatment

1. Introduction

EMG signals display complexity which can be quantified by the fractal theory [1, 2]. Recent research suggests that fractal dimension may be considered as a reliable EMG parameter and is not dependent on the exerted force. [3]. Additionally, fractal analysis has been suggested to have huge potential for wide application in the study of the impact of various stimuli on biosignals, including ECG and EEG and these analyses are very important in physiological and sports sciences paving the way for better understanding of human physiology [1].

In 2013, Ross and Ware published a paper relative to NPT, that was practiced and originated by Ware, based on system dynamics, chaos, and fractal theory [4]. The authors established a theoretical point of view regarding fractal physiology principles for this treatment without elaborating on their experimental data based on surface electromyography (sEMG), electroencephalography (EEG) and heart rate variability (HRV) analyses. The aim of the present paper, which is methodological in principle, is to fill in the experimental gap through the illustration of Ross and Ware into the theoretical treatment by introducing the principles of multifractal theory and by studying the series of experimental data of surface electromyography accompanied by multifractal analysis. The data have been obtained by simultaneously recording the sEMG, the EEG, and the ECG at 1000 Hz of sampling frequency and the analysis of the data has been obtained by performing a multifractal analysis using MATLAB software based on DFA (Detrended Fluctuation Analysis) of the arising time series data. Ten athletes were used for the analyses, all clinically controlled in their fundamental parameters. sEMG, EEG and ECG were performed with the recording of the relative signals at 1000 Hz. In the present paper, we will present the results related to the sEMG. In the following paper, we will deal instead with the results related to the EEG and the R-R signal. We carried out our analysis by means of the Detrended Fluctuation Analysis (DFA) accompanied by the analysis of surrogate data and statistical analysis on the various subjects. A conclusive interpretation of the physiological and clinical profiles is reported.

The purpose of this chapter is twofold. On the one hand, we invite the readers to use multifractal analysis in sEMG teaching them what to do. On the other hand, we suggest the reader to consider the validity of the NPT method that is based on fractal theory and chaos theory and thus it represents a good occasion to study multifractals. NPT is a method that, as explained in the section Theoretical Background, is based on chaos theory, on the multifractality of the temporal sets that emerge from the study. It is shown that the use of multifractality is a very powerful method that can be used safely by the researcher and on the other hand, it is also shown that the indices emerging from these studies, such as the Generalized Hurst Exponents and the other indices that are introduced in this work, and which arrange themselves well during the study, constitute a valid element of analysis and clinical diagnosis and therefore function quite well in every respect.

1.1. Theoretical Background

The historical background of this paper is reported in the article by Ross and Ware [1] which we refer to since it exposes in rigorous detail every clinical aspect regarding NPT treatment on the subject. In particular, the following facts were observed: (1) rapid, involuntary muscle contractions and relaxations, (2) mild and/or vigorous fluctuations in limbs or torso, (3) mild and/or vigorous chaos or periodic patterns in one or more jointed regions, and (4) abrupt changes in activity location, symmetries, velocity, and in-phase/anti-phase relations.

Introducing a new paradigm in the delivery of psychophysical therapy that has relevance to our greater understanding of how the nonlinear dynamical system is perceived and interacts with the environment at highly discrete scales.

For decades, NPT has been attributed to unprecedented highly positive health and wellbeing transitions in very small-time scales for patients who presented with highly complex neurological diseases and disorders deemed untreatable by the present medical paradigm, inclusive of incomplete and complete spinal cord injuries. Other conditions benefiting from NPT include motor neuron disease, multiple sclerosis, Graves' disease, lupus and migraine [1]. The said NPT outcomes have been well documented and publicized by the international and national media. https://www.neurophysicstherapy.global/media/

1.2. The Vital Importance of Scientific Research into NPT Phenomena

It is to be understood that scientific research surrounding NPT is a reverse engineering process. A phenomenon displaying enormously beneficial qualities has been discovered, firstly, and now we need to understand why this is the case from a scientific perspective; That is, what are we observing, how can this new knowledge be transferred towards the greater understanding of what is happening in other nonlinear dynamical systems and do present-day theories surrounding nonlinear dynamical systems need to be expanded upon as a result of this new knowledge.

Observing and measuring a human system during these phases enables the observer to visually appreciate the CNS's relationship with the environment, its perception of the environment, its extreme sensitivity to initial conditions, self-organized criticality, and the ability to recognize the signatures of a system harboring disease or disorder in the absence of knowledge of medical records, when compared to the signatures of a healthy robust system. Obviously, there is a lot of knowledge missing in other fields of the medical and physical sciences, as other fields of medical science have yet to produce such outcomes.

1.3. Systems-Wide Phase Transitions and Their Importance to the Future Health, Wellbeing, and Psychophysical Function of the Human Nervous System

Terms such as phase transitions are generalized terms that summarize a multitude of events taking place in dynamical systems. However, these terms may be used in context to express the sudden shifts in behavior during NPT observed in these systems in phase space.

During all stages of NPT, individuals are perfectly coherent and are able to communicate and start/stop the observed transition dynamics at will, as with all other individuals who have learned how to transcend the levels of normal inhibition of the human system displays to control the underlying chaotic state of the central nervous system - not eliminate chaos, as this is needed to enable us to seamlessly transition between mental states and adapt rapidly to changes in the environment. It is when the brain becomes less chaotic that pathologies such as seizures occur to enable the system to dispose of its excess energy and over-orderliness.

The connection with chaotic theory, chaotic transitions, dynamics of these chaotic transitions, fractality, and multifractality of these EMG systems were also discussed in Ross and Ware's article in detail [4] but a connection between theory and experimental aspects was missing that instead is discussed in the present paper. For the sake of brevity, we refer the reader to this article and other selected bibliographic references of this content [4–23], which we recommend the reader to read very carefully because in this way, he/she will obtain a complete exposition and exhaustive understanding both from a clinical and theoretical point of view. For our present work, we only add that the link between the EMG signals and multifractality is expressed very well in the work of Ross and Ware and we have fully adopted that in the present work. In particular, Ross and Ware expressed the following convictions: The general explanations of the patterns observed above are well established. They are fixed in the nonlinear dynamical nature of complex systems. As one deduces from Haken's theory of non-equilibrium phase transitions, the tendency of such EMG systems is to be in the regime of instability [6]. The loss of stability is outlined by the appearance of new behavioral patterns [7]. Different patterns represent different attractors. These changes are chaotic transitions [6,8] and this means that the result is a new attractor.

The psychophysical system is attracted to certain states to innately maintained patterns of behavior that have a desirable amount of degrees of freedom in the best interest of survival and adaptation to changes within the environments. The innate system has no preference or bias so when there is a change in behavior associated with a disease or disorder then these behaviors are maintained within a rogue metabolic attractor that has limited degrees of freedom. When these patients' systems expose their underlying kinetics, these kinetics display highly deterministic and very regular periodicity. They appear to be stuck in this rouge attractor that can be dimensionally and metabolically measured and are visually constructed via the EMG signals. These systems need a prescribed perturbation which occurs via the NPT process to input more energy and information into the system, so that the system can build up the escape velocity needed to pull out of the rogue attractor, and this would be displayed as phase transitions where the system will then display systems wide flowing movements similarly to flocks of birds flying in unison. This is what occurs and the system exhibits rapid enhancements to function and performance when this occurs. A healthy person's system will display these flowing movements all the time.

The studies of multifractals and chaos are linked in many ways. They share common ideas and methods of analysis. From a methodological point of view, we argue that a systems phase space (set) may be formed from the time series data. When the phase space system is fractal the system that generated the time series is chaotic. The fractal time dimension of the phase space informs us of the minimum number of independent variables that are required to generate the time series. Chaos indicates that the output of a deterministic nonlinear process is so complex that it mimics the random behavior, the point being that chaos analysis is to determine if the experimental data are due to a deterministic or random process and, if this is due to a deterministic process to determine the mathematical kind of that process. The essential properties of a chaotic system are that it is deterministic dynamical system, it is described either by difference or differential equations, it has a strange attractor where, as noted previously, the technical jargon word chaotic means that the system has sensitivity to initial conditions and the word strange means the attractor is fractal. Such fractal dimension is related to the number of independent variables that are needed to generate the time series of the values of variables, it has bifurcations that happen when the form of the trajectory in the phase space changes a parameter passing through a certain value, control that means that small changes in such parameters produce small changes in the subsequent values of the variables. In nonlinear systems, small changes in such parameters produce enormous changes in the subsequent values of the variables because of the dependence to the sensitivity to the initial conditions.

All the previous statements are theoretical elaborations designed in the paper mentioned above [4]. In the present paper, we give indication that the experimental time series, data recorded by sEMG on the examined athletes, are multifractals.

2. Materials and Methods

2.1. Participants

Ten subjects who were male athletes participated in the research. The age of the subjects varied from 19 to 23 years old. Their BMI varied between 22- 25 and their height was between 175 and 180 cm.

The training of such athletes consisted in continuous workouts and Nutritional and Fitness programs supporting such athletes to improve their skills.

Trained athletes were confirmed by ECG, the R-R signal, by the BMI performed by the impedance meter and by lipid echography. All such measurements were performed by our group in addition to other exams performed by our clinicians. All the athletes were submitted to polygraphy analysis.

All the athletes signed an informed consent for the treatment. The health conditions of the subjects were constantly monitored by a group of clinicians during the rest, the treatment by NPT, and the post-treatment phases to evaluate the subjective conditions of the athletes. For at rest, we intend the athletes with muscles at rest but with their autonomous system activated and at alert and thus always ready to be activated, in order to respond to the external stimuli.

2.2. sEMG Recordings and Analysis

All sEMG signals were simultaneously recorded using BioRadio®150 technology (Great Lakes Neurotechnologies, Cleveland, OH, USA) samples at 1000 Hz for the frequency of recording. Such analysis was performed at rest, during NPT and post NPT. The following muscles were selected for the sEMG: right and left trapezius, right and left abdominals, and right and left adductors of the lower limbs. Recordings were also taken during NPT while lying on the floor. A time of 5 minutes was selected for at rest treatment and post treatment analysis.

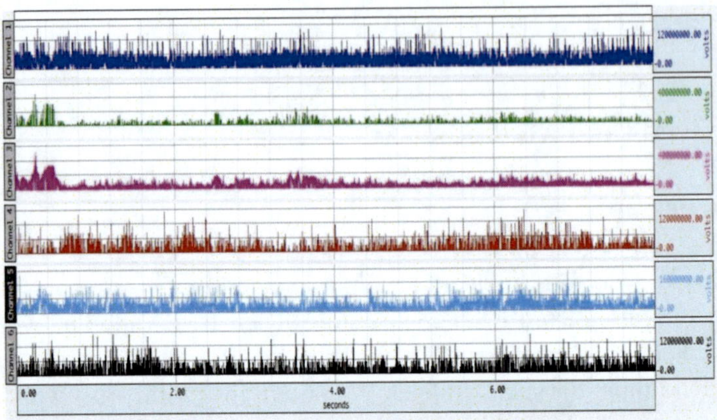

Figure 1a. The recording of sEMG for right trapezius for a time of about 8 seconds. Clinical condition: At rest.

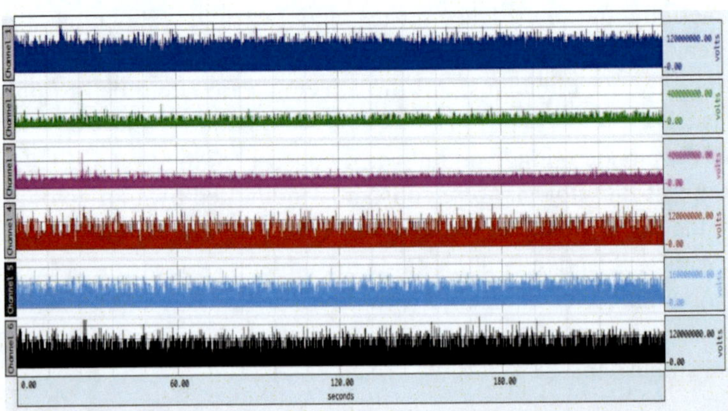

Figure 1b. The recording of sEMG for right trapezius for a time of about180 seconds. Clinical condition: At rest.

The examinations consisted of recording 8 signals arising from six sEMG channels (see Figures 1-3 for the relative signals that relate the recording of the signal at rest before the treatment, during the treatment and post the treatment. Each figure relates the recording of the sEMG signals for the time period of 8 and 200 seconds). The figures are reported in Figures 1a,b; 2a,b;3a,b.

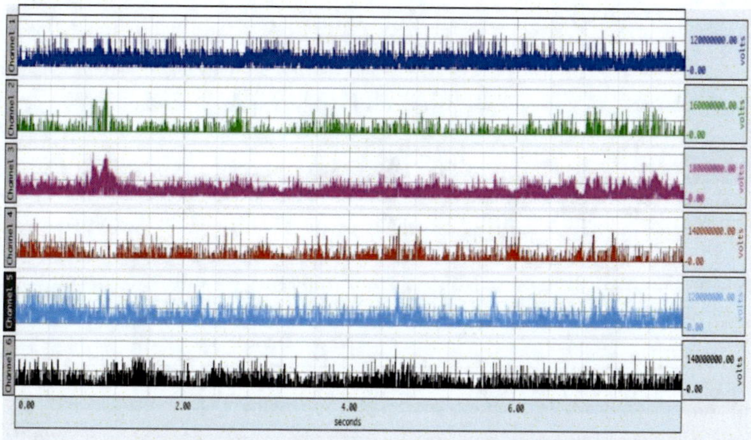

Figure 2a. The recording of sEMG for right trapezius for a time of about 8 seconds. Clinical condition: During NPT.

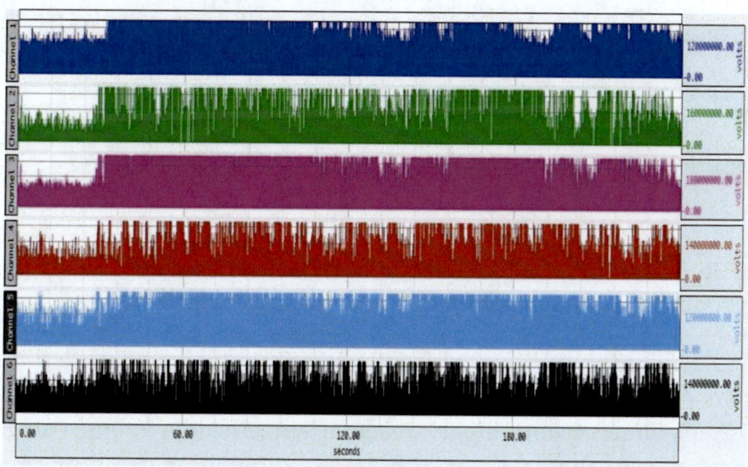

Figure 2b. The recording of sEMG for right trapezius for a time of about 180 seconds. Clinical condition: During NPT.

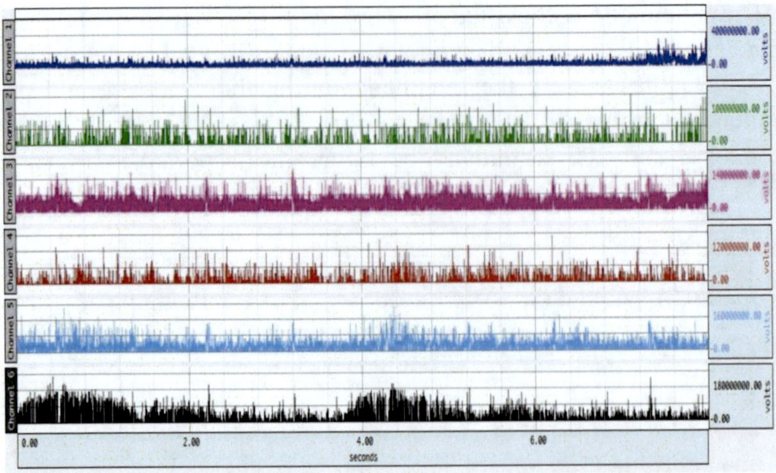

Figure 3a. The recording of sEMG for right trapezius for a time of about 8 seconds. Clinical condition: Post-NPT.

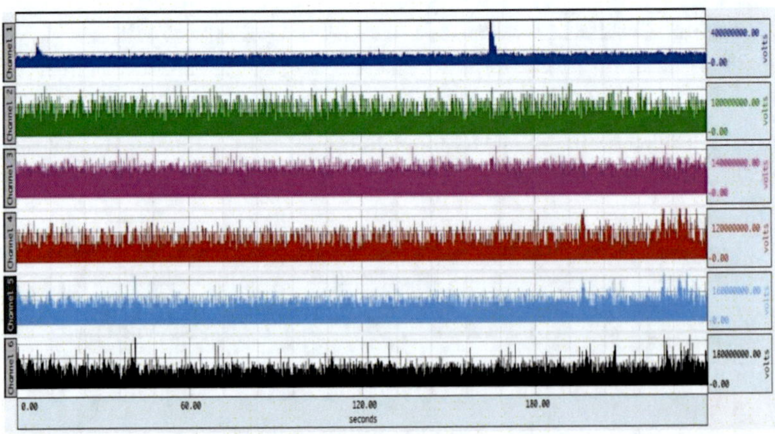

Figure 3b. The recording of sEMG for right trapezius for a time of about 180 seconds. Clinical condition: Post-NPT.

The electrodes were placed lengthwise on the right and the left trapezius for channels 1 (Ch1) and 2 (Ch2), respectively. These regions were selected because the trapezius is highly reactive to activation of the sympathetic nervous system (SNS). Any significant differences between tone and synchronization between the left and right trapezius are indicative of undesirable hemisphere dominance whereas dominantly stable bilateral synchronization is observed in healthy men and women [24, 25].

For the EMG recording of channels 3 (Ch3) and 4 (Ch4), the electrodes were placed on the right and the left abdominals, respectively, bilaterally adjacent to the navel.

The electrodes were placed lengthwise on the adductors of the lower right and the lefty limbs (legs) for channels 5 (Ch5) and 6 (Ch6), respectively.

Placing the EMG electrodes on these three muscles (trapezius, abdominals, and adductors), enabled to monitor the overall system responses.

We used raw data. The sample frequency was 1000,00 Hz and the analysis was extended on the whole 5 minutes signal. Subsequently, each time series arising from one of the six channels was elaborated in the MATLAB software for the Detrended Fluctuations Analysis as developed by Wang et al. [26] and the indices h(q), τ(q), α(q), f(α), D(q) were estimated.

2.3. How to Perform the Methodological Analysis: Theoretical Elaboration

All the arising sEMG time series data were stored on a computer. The procedure of the analysis of the data was developed via a multifractal analysis (MF-DFA 1-D) following the method indicated by Wang et al. [26]. We have six steps of analysis:

Step 1

We calculated the function

$$y(k) = \sum_{t=1}^{k} \left[x(t) - \bar{x}(t) \right], k = 1,2, \dots \dots, N$$

of the fluctuations between the experimental time series, $x(t)$ and its mean value $\bar{x}(t)$.

Step 2

The function $y(k)$ was divided in $N_s = N/s$ not overlapping segments all having the same length s.

Step 3

The local trend $y_v(i)$ was estimated by using the least squares method for each segment.

Since the detrending of each temporal series was obtained by subtracting the polynomial fits from $y(i)$, we estimated $y_v(i)$ by the relation

$$y_s(i) = y(i) - y_v(i)$$

Step 4

We evaluated the "local trend" for each of the $2N_s$ segments by using the least square fit of the series. This is a procedure that finds the best fit behavior respect to a series of points by minimizing the sum of the residuals. The variance was determined in the following manner

$$F^2(s,v) = \frac{1}{s}\sum_{i=1}^{s} y_s^2[(v-1)s+i]\,, v = 1,2,\dots\dots, N_s$$

with

$$F^2(s,v) = \frac{1}{s}\sum_{i=1}^{s} y_s^2[N-(v-N_s)s+i]$$

for

$$v = N_S+1, 2, \dots .2N_S$$

Step 5

By executing the mean on all the segments, we obtained the function of the fluctuations of order q

$$F_q(s) = \left[\frac{1}{2N_s}\sum_{v=1}^{2N_s}[(F^2(s,v))^{q/2}]\right]^{1/q}$$

with q assuming any value except zero.

For $q = 0$, we have instead

$$F_0(s) = exp\left[\frac{1}{4N_s}\sum_{v=1}^{2N_s} ln[F^2(s,v)]\right]$$

Such generalized functions of fluctuations, $F_q(s)$, are q dependent and are correlated with s for various q. This objective is reached by repeating steps 2-5 for the different time scales s.

Step 6

We determined the “scaling behavior” of functions of fluctuations by analysis of the behavior $Ln\ F_q(s)$ vs $Ln\ s$ for any q. In cases where the given time series $x(t)$ results are correlated in accordance with a *“long range”* law of power, $F_q(s)$ will increase by a power law at increasing values of s,

$$F_q(s) \propto s^{h(q)}$$

These are the six steps of the work. Furthermore, due to its affinity with the Hurst exponent, $h(2)$ is called the generalize Hurst exponent. Generally speaking, we have a nonlinear behavior as function of q.

Other fundamental relations are

$$\tau(q) = qh(q) - 1$$

the generalized multifractal dimension

$$D(q) = \frac{\tau(q)}{q-1}$$

The Lipshits-Holder exponent

$$\alpha(q) = \tau'(q)$$

and the multifractal spectrum

$$f(\alpha) = q\alpha(q) - \tau(q)$$

This last observation completes the method that we developed by MATLAB. The index we used are given below.

$$\Delta h = h(q_{min}) - h(q_{max})$$

Since richer multifractality corresponds to higher variability of h(q), the degree of multifractality is quantified by this index.

In addition, we have

$$\Delta\alpha = \alpha_{max} - \alpha_{min}$$

$$\Delta\alpha_L = \alpha_0 - \alpha_{min}$$

$$\Delta\alpha_R = \alpha_0 - \alpha_{max}$$

where α_0 is the particular value of α corresponding to $q = 0$

$$\Delta f(\alpha) = f(\alpha_{min}) - f(\alpha_{max})$$

$$f(\alpha) = A(\alpha - \alpha_{max})^2 + B(\alpha - \alpha_{max}) + C$$

with A, B, C being the coefficients of the parable.

The other step is that one of the surrogate analysis that is able to convince us that we are in presence of a multifractal. This Surrogate Data Analysis (shuffled data) was used to test and to verify the multifractality of the experimental data and consisted in the analysis of $F_q(s)$, the Average Mutual Information and the multifractal spectrum $f(\alpha)$ of our data (see the Figures 4, 5, 6, 7).

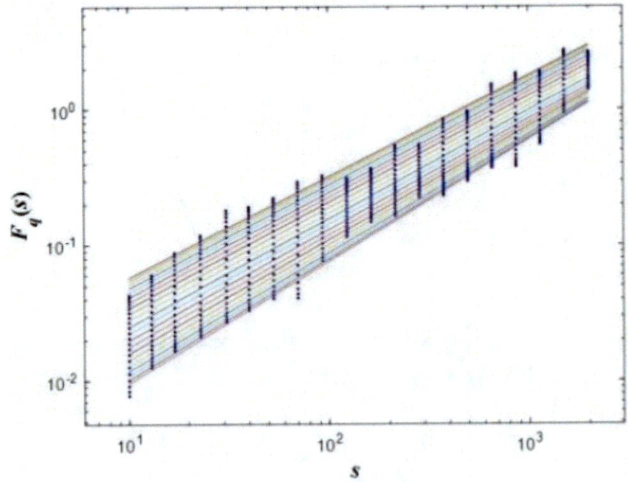

Figure 4. We report $F_q(s)$ against s to convince us that we are in presence of a multifractal.

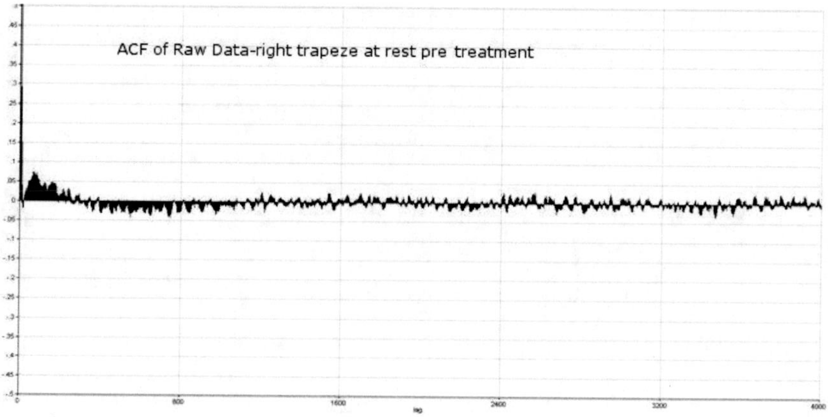

Figure 5. ACF of the shuffled data.

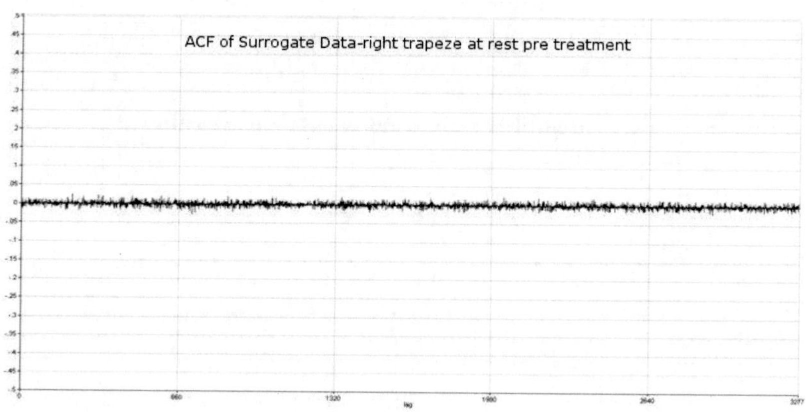

Figure 6. ACF of the shuffled data. The results confirm that we are in presence of a multifractal.

2.4. Statistical Analysis

After the proper verification for normality and sphericity, the Repeated-Measures ANOVA statistical analysis was used to compare the data of pre-, during- and post-NPT for each electrode. A p value less than 0.05 was considered as statistically significant.

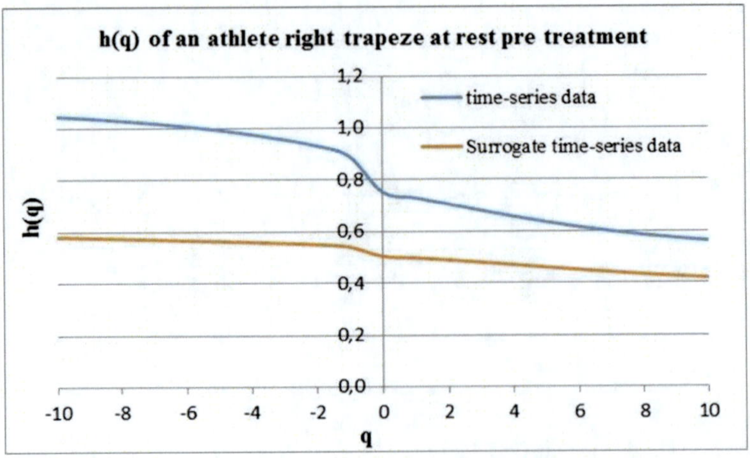

Figure 7. These results confirm that we are in presence of a multifractal.

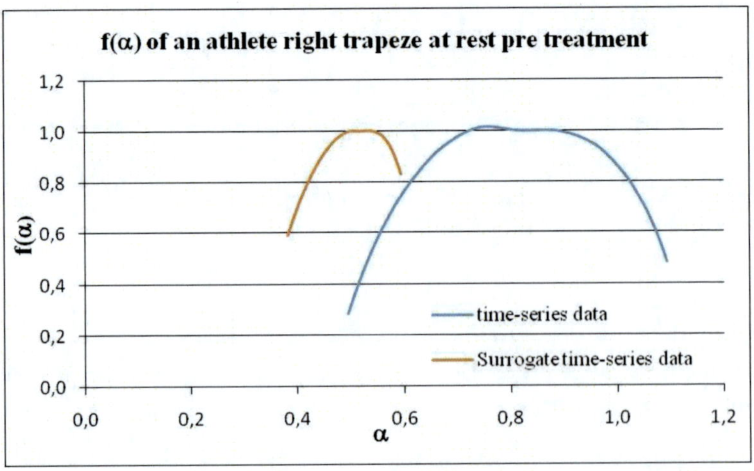

Figure 8. These results confirm that we are in presence of a multifractal.

3. Results

3.1. The Multifractal Analysis

For the multifractal analysis, the calculated indices h(q), D(q),τ(q), α(q), f(α), are shown in the Figures 9-13.

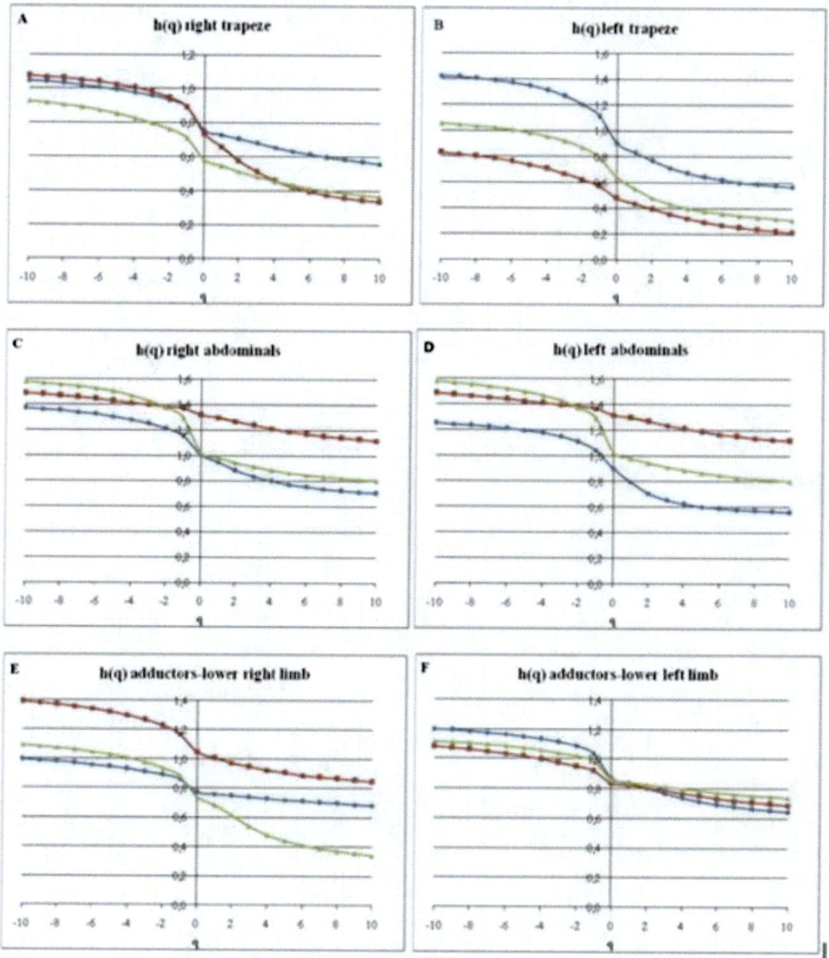

Figure 9. Representation of h(q) as function of q: The right (A) and the left (B) trapezius, the right (C) and left (D) abdominals and the right (E) and left (F) adductors of athletes at rest pre-( ), during-( ) and post-( )-treatment (at rest) are shown. We have the typical behavior of the multifractals dynamics in all the examined phases of pre-, during- and post-treatment: the whole of the exponents h(q) describes the scaling of the qth-order fluctuation function. Multifractality relates the scaling property of a time series that is represented by an array of scaling exponents, and not by any single one. The multifractality of a time series is identified by examining the dependence of h(q) on high and low-order moments. If the time series is monofractal, then h(q) is independent of q. If the series is multifractal, then h(q) depends on q, characterizing in this manner a single form of self-similarity over time. h(q) depends significantly on q only if large and small fluctuations scale differently. With major or minus difference, we find such a situation in the pre-, during- and post-treatment in all the examined cases.

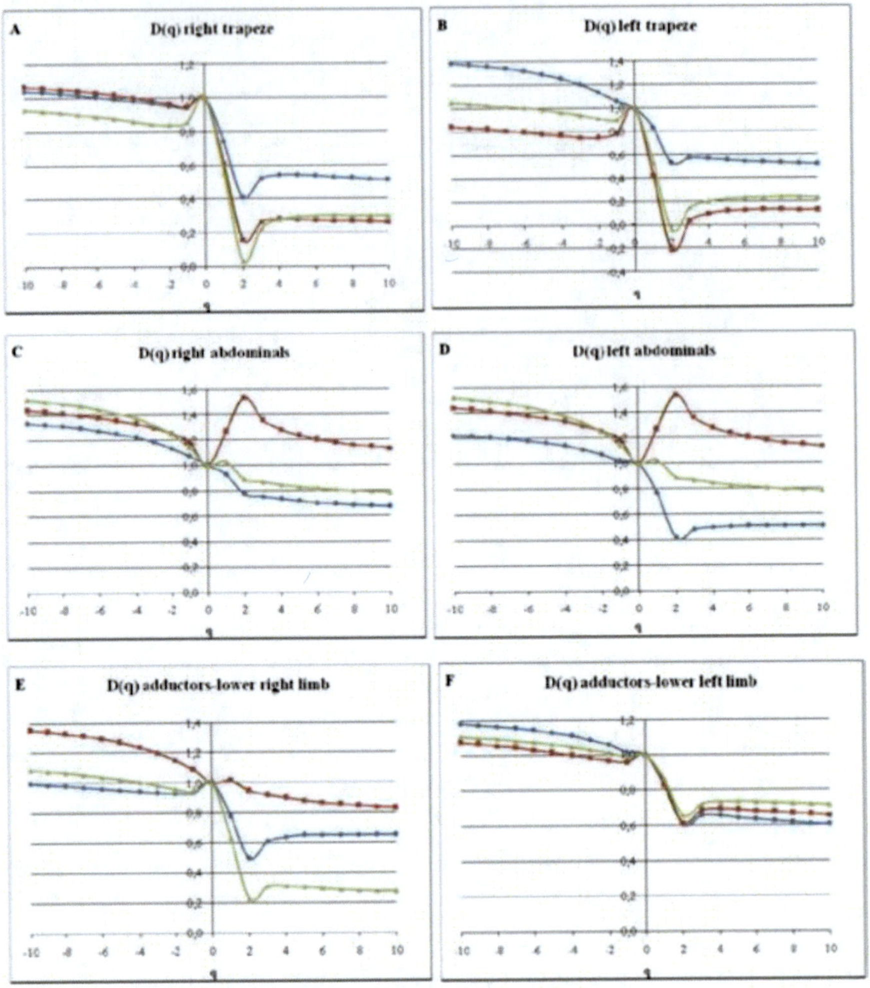

Figure 10. Representation of D(q) as function of q: The right (A) and the left (B) trapezius, the right (C) and left (D) abdominals and the right (E) and left (F) adductors of athletes at rest pre-( ), during-( ) and post-( )-treatment (at rest) are shown. Generally speaking, D(q) is a measure of the inhomogeneity or multifractality of an attractor whenever the D(q) values are not equal for all q. However, when the D(q) are the same, the distribution of vector points on the attractor is fairly homogeneous.
A chaotic attractor with non-constant values of the generalized dimension is called a multifractal. The results in this picture are therefore in line with the interpretation that signals of sEMG as we calculated them for trapezius, abdominals and adductors are time series of a multifractal.

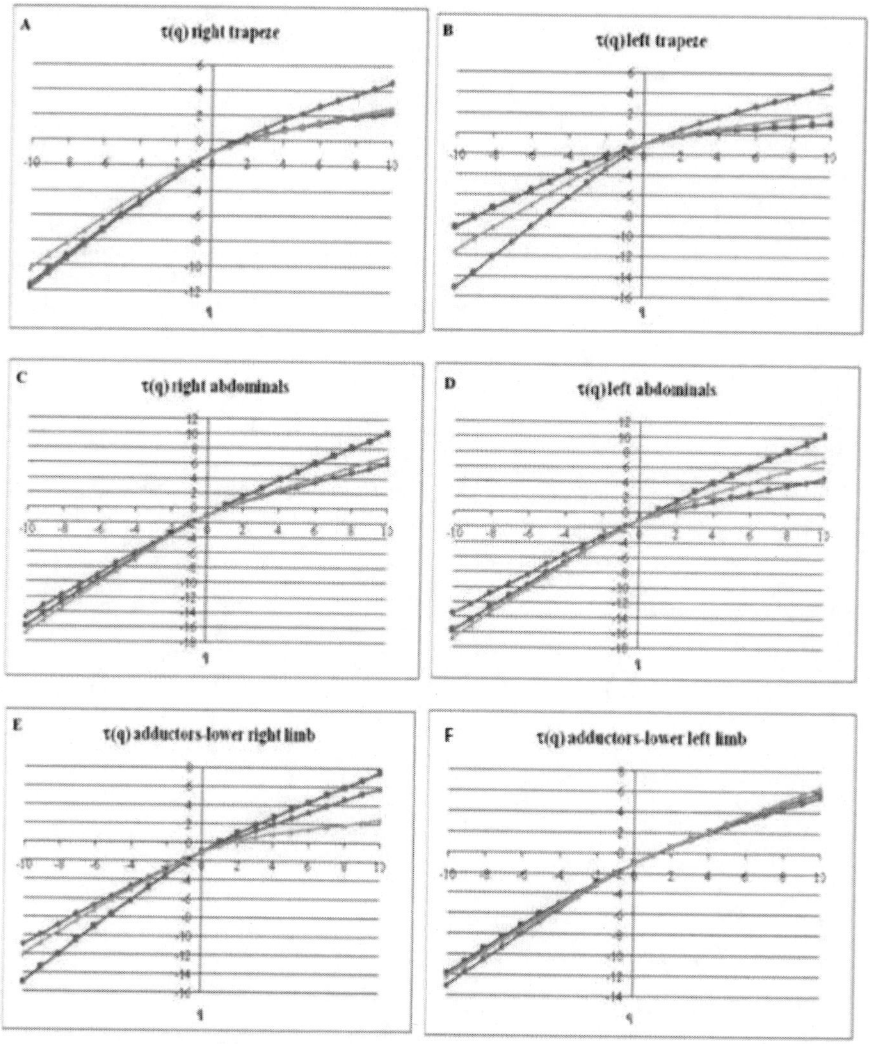

Figure 11. Representation of $\tau(q)$ as function of q: The right (A) and the left (B) trapezius, the right (C) and left (D) abdominals and the right (E) and left (F) adductors of athletes at rest pre-( ), during-( ) and post-( )-treatment (at rest) are shown. The h(q) obtained using MFDFA is related to the Rényi exponent $\tau(q)$ by the relation, previously introduced, $\tau(q)= qh(q)-1$.

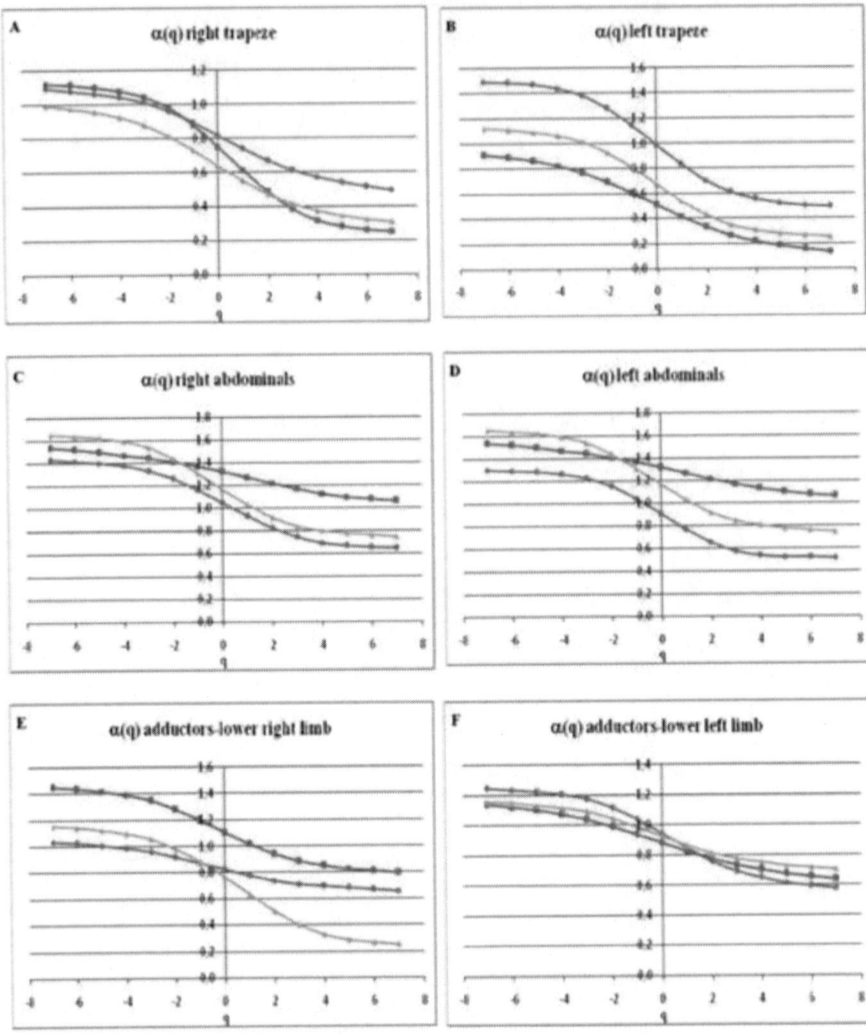

Figure 12. Representation of α(q) as function of q: The right (A) and the left (B) trapezius, the right (C) and left (D) abdominals and the right (E) and left (F) adductors of athletes at rest pre-( ), during-( ) and post-( )-treatment (at rest) are shown. α is the Hölder exponent or singularity strength, and it gives the multifractal spectrum *f*(α), which describes multifractal time series. For pure multifractals, the value of α decreases with increased fluctuation as it is obtained in these graphs for right and left trapezius, abdominals and adductors in condition at rest, pre- and post-treatment, as well as during the treatment.

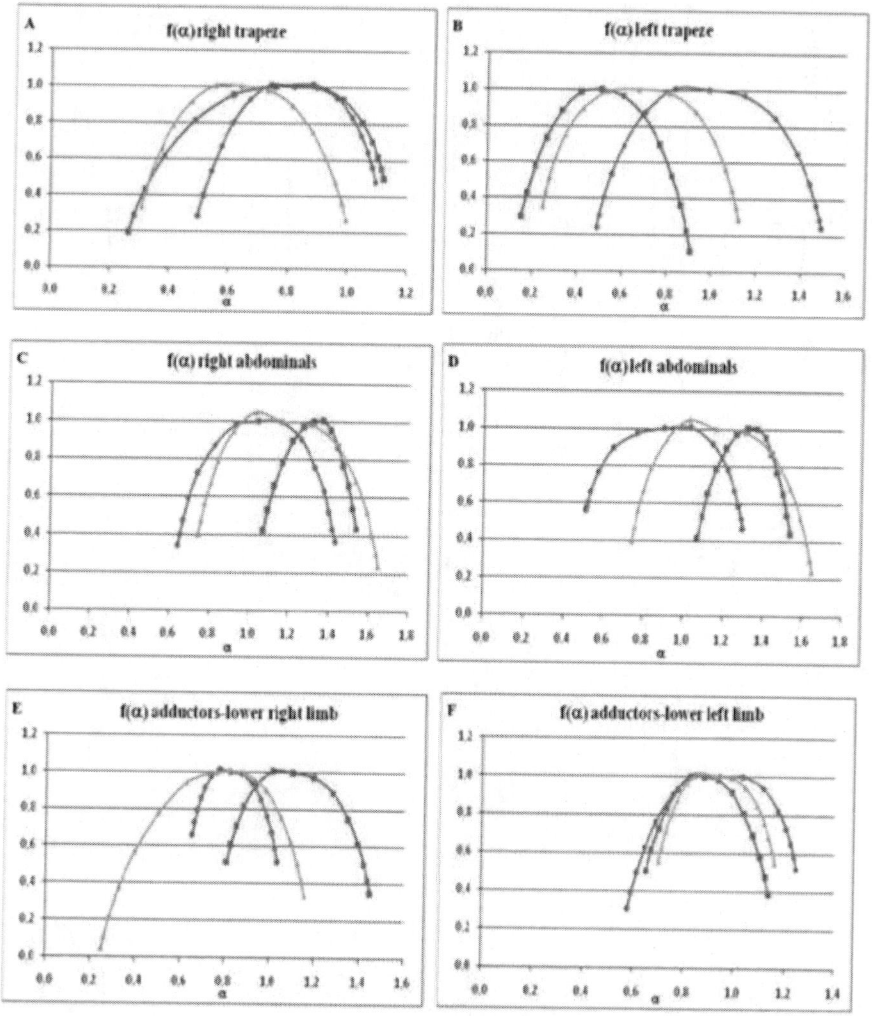

Figure 13. Representation of $f(\alpha)$, the multifractal spectrum, as function of α: The right (A) and the left (B) trapezius, the right (C) and left (D) abdominals and the right (E) and left (F) adductors of athletes at rest pre-( ), during-( ) and post-( )-treatment (at rest) are shown. In all cases, the multifractal spectra are quite asymmetrical with convex parables, which confirms that the time series for the six selected regions are multifractal, but to a varying degree, since the curves are all different. All of them reach the maximum value $f(\alpha)$ of 1, relating to the single dimension of the studied variables. When the spectra are compared, we observe that their shapes are different; consequently, all of the studied data sets are subjected to different high and low fluctuations.

Table 1. Mean values ± standard deviation of the generalized hurst exponent, h(2), $\Delta\alpha$, Δf, the coefficients A, B, C, Δh, $\Delta\alpha_L$, $\Delta\alpha_R$, recorded by channels 1 (right trapezius) 2 (left trapezius), 3 (right abdominals), 4 (left abdominals), 5 (right adductors) and 6 (left adductors) pre-NPT treatment (color white), during NPT treatment (color green) and post-NPT treatment (color pink) at rest in the healthy male subjects (n = 10)

	H(2)	$\Delta\alpha$	Δf	Δh	$\Delta\alpha_L$	$\Delta\alpha_R$	A	B	C
CHANNEL 1 (right trapezius)									
Pre-treatment	0,796 ± 0,084	0,512 ± 0,059	-0,336 ± 0,224	0,413 ± 0,050	0,304 ± 0,039	-0,207 ± 0,055	-8,281 ± 1,182	-3,472 ± 0,880	0,674 ± 0,161
During treatment	0,551 ± 0,026	0,696 ± 0,002	-0,375 ± 0,164	0,560 ± 0,023	0,412 ± 0,034	-0,284 ± 0,036	-5,798 ± 0,557	-3,418 ± 0,058	0,534 ± 0,004
Post-treatment	0,472 ± 0,047	0,846 ± 0,163	-0,051 ± 0,123	0,725 ± 0,169	0,439 ± 0,115	-0,407 ± 0,049	-4,064 ± 1,782	-3,200 ± 0,982	0,444 ± 0,134
CHANNEL 2 (left trapezius)									
Pre-treatment	0,669 ± 0,129	0,694 ± 0,169	0,175 ± 0,038	0,413 ± 0,050	0,304 ± 0,039	-0,207 ± 0,055	-4,212 ± 1,379	-3,024 ± 0,350	0,511 ± 0,040
During treatment	0,546 ± 0,160	0,444 ± 0,227	-0,149 ± 0,122	0,345 ± 0,193	0,237 ± 0,139	-0,207 ± 0,088	-14,66 ± 8,432	-4,991 ± 1,236	0,555 ± 0,075
Post-treatment	0,589 ± 0,167	0,957 ± 0,137	-0,076 ± 0,476	0,825 ± 0,099	0,503 ± 0,157	-0,454 ± 0,022	-2,894 ± 0,310	-2,692 ± 0,432	0,436 ± 0,093
CHANNEL 3 (right abdominals)									
Pre-treatment	0,947 ± 0,076	0,905 ± 0,267	0,103 ± 0,136	0,787 ± 0,241	0,434 ± 0,141	-0,471 ± 0,130	-3,518 ± 1,269	-3,078 ± 0,275	0,383 ± 0,103
During treatment	0,908 ± 0,090	0,742 ± 0,183	-0,079 ± 0,210	0,625 ± 0,151	0,385 ± 0,079	-0,357 ± 0,116	-4,831 ± 1,313	-3,240 ± 0,389	0,477 ± 0,223
Post-treatment	0,849 ± 0,133	1,034 ± 0,183	-0,007 ± 0,235	0,882 ± 0,151	0,512 ± 0,132	-0,522 ± 0,052	-3,166 ± 0,469	-3,260 ± 0,152	0,267 ± 0,031

	H(2)	$\Delta\alpha$	Δf	Δh	$\Delta\alpha_L$	$\Delta\alpha_R$	A	B	C
CHANNEL 4 (left abdominals)									
Pre-treatment	0,970 ± 0,036	0,853 ± 0,090	0,208 ± 0,179	0,746 ± 0,074	0,391 ± 0,078	-0,461 ± 0,016	-3,190 ± 0,190	-2,972 ± 0,098	0,391 ± 0,014
During treatment	0,949 ± 0,128	0,479 ± 0,190	0,003 ± 0,147	0,397 ± 0,168	0,246 ± 0,106	-0,234 ± 0,091	-9,725 ± 5,281	-2,434 ± 3,699	0,593 ± 0,095
Post-treatment	0,903 ± 0,039	0,794 ± 0,181	0,043 ± 0,014	0,674 ± 0,161	0,389 ± 0,083	-0,405 ± 0,098	-4,240 ± 1,253	-3,311 ± 0,203	0,396 ± 0,104
CHANNEL 5 (right adductors)									
Pre-treatment	0,872 ± 0,057	0,373 ± 0,031	-0,108 ± 0,461	0,300 ± 0,028	0,200 ± 0,087	-0,174 ± 0,059	-10,93	-2,526 ± 3,934	0,679 ± 0,230
During treatment	0,618 ± 0,116	0,901 ± 0,228	-0,093 ± 0,355	0,765 ± 0,201	0,464 ± 0,169	-0,438 ± 0,063	-3,907 ± 1,194	-3,297 ± 0,626	0,384 ± 0,064
Post-treatment	0,727 ± 0,162	0,840 ± 0,095	-0,574 ± 0,402	0,691 ± 0,085	0,523 ± 0,016	-0,316 ± 0,110	-3,991 ± 0,135	-2,560 ± 0,843	0,590 ± 0,256
CHANNEL 6 (left adductors)									
Pre-treatment	0,873 ± 0,100	0,534 ± 0,187	-0,109 ± 0,147	0,440 ± 0,166	0,280 ± 0,117	-0,255 ± 0,070	-8,495 ± 4,328	-3,984 ± 0,968	0,576 ± 0,009
During treatment	0,733 ± 0,128	0,722 ± 0,174	-0,398 ± 0,231	0,598 ± 0,151	0,426 ± 0,129	-0,296 ± 0,052	-5,402 ± 2,400	-3,069 ± 0,777	0,600 ± 0,053
Post-treatment	0,965 ± 0,257	0,479 ± 0,112	-0,149 ± 0,142	0,373 ± 0,109	0,260 ± 0,048	-0,220 ± 0,076	-9,900 ± 3,262	-4,145 ± 0,073	0,538 ± 0,060

In Table 1, we report the results of the mean values and standard deviation of the generalized Hurst exponents h(2),$\Delta\alpha$, Δf, A, B, C, Δh, $\Delta\alpha_L$, $\Delta\alpha_R$, in the condition of pre-treatment, during the NPT treatment and post- treatment.

The use of the Generalized Hurst Exponents and the other indices represent a particular tool of the multifractal analysis. Concerning our NPT analysis, we observe that their values, in the passage from at rest to NPT exercise, and after returning to the at rest condition, implies a modification of the values of such indices that, therefore, represent fundamental elements to characterize the physiological condition of the investigated muscles. We have the range of normality as they are given by their mean values plus standard deviation. After we have the mean values and standard deviation during the NPT treatment.

Such results indicate the modification of the muscle regime during the NPT and the transitions of such muscle dynamics during the NPT and the importance of using such multifractal indices, in general, in current sEMG studies owing to their sensitivity to the physiological conditions of the muscles. Such indices modify their values during the NPT treatment and after they return to normal values. It is this datum that confirms the effectiveness of our treatment. In fact, the passage of such indices indicates that the NPT is an effective treatment that rearranges the physiological dynamics of the muscle.

$\Delta\alpha_L$ and $\Delta\alpha_R$ evidence a rather constant regime of the fluctuations ranging $\Delta\alpha_L$ from 0.304 ± 0.039 to 0.412 ± 0.034 and to 0.439 ± 0.115 for the right trapezius in the experimental conditions of pre-, during- and post-NPT treatment, passing from 0.304 ± 0.039 to 0.237 ± 0.139 to 0.503 ± 0.157 in the phases pre-, during- and post-NPT treatment for the left trapezius, from 0.434 ± 0.141 to 0.385 ± 0.079, to 0.512 ± 0.132 in the phases of pre-, during- and post- treatment for the right abdominals, to 0.391 ± 0.078 to 0.246 ± 0.106 to 0.389 ± 0.083 in the phases of pre-, during- and post-NPT treatment for the left abdominals, and, finally, for 0.200 ± 0.087 to 0.464 ± 0.169 to 0.523 ± 0.016 in the phase of pre-, during- and post-NPT treatment for the right adductors and 0.280 ± 0.117 to 0.426 ± 0.129 to 0.260 ± 0.048 on the phases of pre-, during- and post-NPT treatment for the left adductors. $\Delta\alpha_R$ vary from - 0.207 ± 0.055 to -0.284 ± 0.036 to -0.407 ± 0.049 in the phases pre-, during- and post-NPT treatment for the right trapezius, from -0.207 ± 0.055, to -0.207 ± 0.088, to -0.454 ± 0.022 in pre-, during- and post-NPT treatment for the left trapezius, from -0.471 ± 0.130 to -0.357 ± 0.116 to -0.522 ± 0.052 in the phases of pre-, during- and post-NPT treatment for the right abdominals and from -0.461 ± 0.016 to -0.234 ± 0.091 to -0.405 ± 0.098 in the phase of pre-, during- and post-NPT treatment for the left abdominals, and from -0.174 ± 0.059, to -0.438 ± 0.063 to -0.316 ± 0.110 in the phases of pre-, during- and post-NPT treatment for the right adductors and from -0.255 ± 0.070, to -0.296 ± 0.052 to -0.220 ± 0.076 in the phases of pre-, during- and post-NPT treatment for the left adductors.

Next, an examination will be made of the behavior of the multifractal spectrum $f(\alpha)$ and the Δf. $f(\alpha)$ was fitted by a parable with the coefficients A, B, C as explained in detail in the text. For the right trapezius we find A = - 8.281 ± 1.182, B = -3.472 ± 0.880, C = 0.674 ± 0.161 in the condition at rest and pre-NPT treatment. This regime evolves through new transitions to A = - 5.798 ± 0.557, B = - 3.418 ± 0.058 and C = 0.534 ± 0.004 during the NPT treatment and to A = - 4.064 ± 1.782, B = - 3.200 ± 0.982, C = 0.444 ± 0.134 in the phase of post-NPT treatment.

Again, the data on $f(\alpha)$ confirms that for the multifractal spectrum, we have a regime of total rearrangement of the condition of multifractality in passing from the pre- to during and post-NPT treatment with statistically significant differences. For the left trapezius, we have A = - 4.212 ± 1.379. B = - 3.024 ± 0.350, C = 0.511 ± 0.040 in the phase of pre-NPT treatment and it becomes A = - 14.666 ± 8.432, B = - 4.991 ± 1.236, C = 0.555 ± 0.075 during the NPT treatment and A= - 2.894 ± 0.310, B = - 2.692 ± 0.432 and C = 0.436 ± 0.093 in the at rest post-NPT treatment. Through this data we confirm again that, by the NPT treatment, we have a total regime of multifractality that is during the pre and post-NPT treatment and during the NPT treatment with transitions of this multifractality regime that is signed from the reported values of A, B, and C.

In the right abdominals, we have A = - 3.518 ± 1.269, B = - 3.078 ± 0.275, C = 0.383 ± 0.103 at rest before the NPT treatment, A = - 4.831 ± 1.313, B = - 3.240 ± 0.389, C = 0.477 ± 0.223 during the NPT treatment and A = - 3.166 ± 0.469, B = -3.260 ± 0.152, C = 0.267 ± 0.031 in the at rest, post the NPT treatment. In the left abdominals, it is instead A = -3.190 ± 0.190, B = - 2.972 ± 0.098, C = 0.391 ± 0.014 in the pre-NPT treatment, A = -9.725 ± 5.281, B = - 2.434 ±3.699, C = 0.593 ± 0.095 during the NPT treatment and A = - 4.240 ± 1.253, B = - 3.311 ± 0.203 and C = 0.396 ± 0.104 in the at rest, post NPT treatment.

For the right adductors, we have instead, for the right, A = -10.930 ± 2.020, B = - 2.526 ± 3.934, C = 0.679 ± 0.230 at rest, pre-NPT treatment, A = - 3.907 ± 1.194, B = - 3.297 ± 0.626, C = 0.384 ± 0.064 during the NPT treatment and A = -3.991 + 0.135, B = -2.560 ± 0.843, C = 0.590 ± 0.256 in the post-NPT treatment.

For the left adductors, we have instead A = - 8.495 ± 4.328, B = - 3.984 ± 0.968, C = 0.576 ± 0.009 in the pre-NPT treatment, A = -5.402 ± 2.400, B = -3.069 ± 0.777, C = 0.600 ± 0.053 during the NPT treatment and A=-9.900 ± 3.262, B = - 4.145 ± 0.073, C = 0.538 ± 0.060 in the regime of the at rest, post-NPT treatment.

Looking to the nonlinear behavior of the curves, we conclude that all the recorded signals exhibit a multifractal regime. In particular, the behavior of f(α), the multifractal spectrum, confirms that we are in the presence of a multifractal regime in all the recorded channels relating the right and left trapezius, the right and left adductors and the right and left abdominal muscles. We pass to the next step, the surrogate data analysis. We calculated $F_q(s)$ against s and we performed the Surrogate Data Analysis by calculating the AutoCorrelation Function (ACF), h(q) against q and the multifractal spectrum of the data and their Surrogate. The results are reported in Figures 4-8.

Let us take a look at the statistical analysis. By this statistical analysis, we have significant results for h(2) for the left trapezius. For Δα, we have significant results for the left trapezius and the left abdominals. For Δf, we obtain significant results for the right trapezius. For Δh, significant results are obtained for the left trapezius and the left abdominals. ΔaL as well as ΔaR give statistical significance for the left abdominals. The coefficients A, B and C are statistically significant for the left trapezius and left abdominals. The level of significance is reported in Table 2 with stars.

The outcome and significance of the used indices become self-evident, if we think that the athletes obtained a significant rearrangement of such indices during the NPT treatment with a subsequent return to their normal conditions after the NPT Treatment. The datum evidences, consequently, that the NPT treatment modifies such parameters with a given significance and that they return to be normal after the NPT Treatment.

4. Discussion

In this section we give a look at the h(2) and $\Delta h = h(q_{min})-h(q_{max})$ values. The human body system needs to be functioning in a coordinated manner where no privileged point of observation is able to be observed. This means that regardless of where you look at the system and what instrument you measure the system with, it should likely be similar across all scales. Energy and information need to be evenly distributed to all parts of the system. In the case of a disease or disorder, this is not the case.

Table 2. Results of statistical analysis, repeated-measures ANOVA, for h(2), $\Delta\alpha$, Δf, A, B, C, Δh, $\Delta\alpha_L$, $\Delta\alpha_R$ comparing the indices in the conditions of pre-, during- and post-treatment

	H(2)	Δα	Δf	Δh	ΔαL	ΔαR	A-B-C
CH1 (right trapezius)							
P value	1	1	0,0260	1	1	1	1
P value summary	ns	ns	*	ns	ns	ns	ns
P < 0.05	No	No	Yes	No	No	No	No
F	-2,1610	-2,2720	10,4100	-2,3330	-2,2330	-2,4440	-2,2420
R squared	13,4200	8,3550	0,8388	7,0120	9,5880	5,5100	9,2590
CH2 (left trapezius)							
P value	P<0.0001	0,0104	0,8057	0,0049	0,0709	0,0021	0,0297
P value summary	***	*	Ns	**	ns	**	*
P < 0.05	Yes	Yes	No	Yes	No	Yes	Yes
F	28,8500	7,4590	0,2208	9,4850	3,4890	12,1800	5,1030
R squared	0,8523	0,5987	0,04228	0,6548	0,4110	0,7089	0,5051
CH3 (right abdominals)							
P value	1	1	0,3724	1	1	1	1
P value summary	ns	ns	Ns	ns	ns	ns	ns
P < 0.05	No	No	No	No	No	No	No
F	-9,2270	-3,7840	1,1700	-3,8150	-3,7080	-3,9780	-3,6700
R squared	2,1830	4,8290	0,2806	4,6810	5,2370	4,0680	5,4780
CH4 (left abdominals)							
P value	1	0,0096	0,1557	0,0114	0,0165	0,0084	0,0356
F value summary	ns	**	ns	*	*	**	*
P < 0.05	No	Yes	No	Yes	Yes	Yes	Yes
F	-7,6530	8,7910	2,3670	8,2390	7,1590	9,2180	5,2070
R squared	2,0950	0,6873	0,3718	0,6732	0,6415	0,6974	0,5655

Table 2. (Continued)

	H(2)	Δα	Δf	Δh	ΔαL	ΔαR	A-B-C
CH5 (right adductors)							
P value	1	1	0,4633	1	1	1	1
P value summary	ns	ns	ns	ns	ns	ns	ns
P < 0.05	No	No	No	No	No	No	No
F	-4,2500	-1,0870	1,1580	-1,0950	-1,1030	-1,1360	-2,8340
R squared	3,4000	12,4900	0,5367	11,5700	10,7100	8,3300	3,3980
CH6 (left adductors)							
P value	1	1	1	1	1	1	1
P value summary	ns	ns	ns	ns	ns	ns	ns
P < 0.05	No	No	No	No	No	No	No
F	-2,4670	-4,1250	-1,7070	-3,2920	-3,2820	-8,0990	-3,6690
R squared	5,2810	3,6670	2,4150	11,2800	11,6500	2,6140	2,1980

The raw sEMG data harvested from the trapeze, abdominals and adductors should all look the same or very similar so that when an event occurs at one location the same event can be observed at the other locations at the same time scale simultaneously. When this level of synchronicity is not being observed in the time series, then this is indicative of a system harbouring a disease or disorder (even without knowledge of patients' medical history). However, as the patient engages with the NPT process in the following days, the system displays much more deterministic behaviors of motor units across all scales (channels) as visually evidenced in their time series portraits in the absence of analyses.

The generalized Hurst Exponent h(2), and $\Delta h = h(q_{min})\text{-}h(q_{max})$ will first be discussed in this section. Different q values have varying effects on the fluctuation functions. The positive q segment v with large variance describes the scaling behavior of the segments with large fluctuations and, in general, large fluctuations are characterized by a smaller scaling exponent for multifractal time series. For negative q, segments ν with small variance will dominate the average F(ν). Thus, if q is negative, the scaling exponent describes the scaling behavior of segments with small fluctuations, usually characterized by a larger scaling exponent.

In Table 1, we see that we have values of h(2) greater than 0.5 in the case of pre-treatment at rest and during NPT for the Ch1 (right trapezius) and the results are greater than 0.5 in pre-, during- and post-treatment in channels ch2 (left trapezius), ch3 (right abdominals), ch4 (left abdominals), ch5 (right adductors), and ch6 (left adductors). For ch1 and ch2, the values of h(2) are about 0.5 during the NPT treatment and are less of 0.5 in the post-NPT treatment. Values greater than 0.5 indicate that the time series have long-term, correlated and persistent fluctuations. Conversely, values of h(2) that are less than 0.5 indicate that we are in presence of a time-series in which the fluctuations are anti-persistent. Values of h(2) around 0.5 instead indicate a completely uncorrelated series. However, the value applicable to series for which the autocorrelations at small time lags can be positive or negative but where the absolute values of the autocorrelations decay exponentially quickly to zero.

In conclusion, starting with the analysis of the generalized Hurst exponent h(2) and $\Delta h = h(q_{min})\text{-}h(q_{max})$, we can conclude that there are modifications of the multifractal regime during the NPT treatment respect to the pre and post treatment with a drastic change of the multifractal regime that passes from the stationary, correlated and persistent fluctuations in the phase of at rest pre-treatment to the phase of uncorrelated signal in the phase of the NPT treatment to the regime of short term, non-stationary, intermittent and non-persistent behavior during post-NPT treatment for the cases of trapezius channels (Ch1 and Ch2). We pass from the value of $h(2) = 0.796 \pm 0.084$ for the pre-treatment to the value of $h(2) = 0.551 \pm 0.026$ during the NPT treatment and to $h(2) = 0.472 \pm 0.047$ for the post-treatment for right trapezius and to the values of $h(2) = 0.669 \pm 0.129$ in the phase of the pre-treatment to the value of $h(2) = 0.546 \pm 0.160$ during the treatment and to the value $h(2) = 0.589 \pm 0.167$ for the post- treatment for the left trapezius. For the other channels relating the abdominals and the adductors (the other channels Ch3, Ch4, Ch5 and Ch6) for the transitions we move in the direction of the stationary, correlated, and persistent fluctuations but with different values of h(2) and Δh in the case of pre-NPT treatment respect to the situation of during and post-NPT treatment.

With regard to α, $\Delta\alpha$, $\Delta\alpha_R$, and $\Delta\alpha_L$, the behavior of alpha is nonlinear in all the case examined of alpha pre-, during- and post-NPT treatment in the trapezius, adductors, and abdominals. The width $\Delta\alpha$ of a multifractal spectrum provides a measure of multifractality: a larger range of $\Delta\alpha$ indicates a wider multifractal spectrum, implying a higher degree of multifractality. A larger or smaller multifractal signal (corresponding to a larger or smaller $\Delta\alpha$) implies a greater or lesser heterogeneous signal. A signal is heterogeneous if it is characterized by sudden bursts of high frequency, intermittencies and/or irregularities, respectively. $\Delta\alpha_L$ and $\Delta\alpha_R$ are the left- and right-hand branches of the multifractal spectrum curve; their values describe the distribution patterns of high and low fluctuations, respectively [27,28]. The asymmetry index (R), which ranges from -1 to 1, quantifies the deviations of the multifractal spectrum curve [29].

For the right trapezius, we have $\Delta\alpha = 0.512 \pm 0.059$ in the phase of pre-NPT treatment, $\Delta\alpha = 0.696 \pm 0.002$ during the NPT treatment and $\Delta\alpha = 0.846 \pm 0.163$ in the phase of post-NPT treatment. In substance we obtain an increment of multifractality passing from pre to the post- treatment.

The behavior is different for the left trapezius, passing from $\Delta\alpha = 0.694 \pm 0.169$ in the pre-treatment test to $\Delta\alpha = 0.444 \pm 0.227$ during the NPT treatment and $\Delta\alpha = 0.957 \pm 0.137$ in the phase of post-NPT treatment. Indeed, in the right trapezius, we have an increase of multifractality in passing from the pre-, to during- and to post- treatment, while instead in the left trapezius, we have a reduction of multifractality in passing from the pre-, to during- the treatment and increasing multifractality in the regime of post-NPT treatment.

For the right abdominals, we have instead $\Delta\alpha = 0.905 \pm 0.267$ in the phase of pre-NPT treatment, to $\Delta\alpha = 0.742 \pm 0.183$ during the NPT treatment and $\Delta\alpha= 1.034 \pm 0.183$ in the post-NPT treatment. We have a reduction of multifractality passing from pre to during the NPT treatment and an increase of multifractality passing from the pre or during NPT treatment to post-NPT treatment. For the left abdominals, we have $\Delta\alpha = 0.853 \pm 0.090$ in the pre-NPT treatment, $\Delta\alpha = 0.479 \pm 0.190$ during the treatment and $\Delta\alpha = 0.794 \pm 0.181$ in the post-NPT treatment. We obtain again a diminution of the multifractals in passing from the pre to during the NPT treatment and an increase in multifractality going from the phase of during the NPT treatment to the regime of post the NPT treatment.

For the right adductors, we have $\Delta\alpha = 0.373 \pm 0.031$ for the pre-treatment, $\Delta\alpha = 0.901 \pm 0.228$ during the NPT treatment and $\Delta\alpha =0.840 \pm 0.095$ during the post- treatment.

The left adductors show instead $\Delta\alpha = 0.534 \pm 0.187$ in the pre-treatment condition, $\Delta\alpha = 0.722 \pm 0.174$ during the NPT treatment and $\Delta\alpha = 0.479 \pm 0.112$ in the post-NPT treatment. For the right and left adductors, we have an increase of multifractality in passing from the pre-NPT treatment to during the NPT treatment and a reduction in passing from the during the NPT treatment to the post-NPT treatment.

In conclusion, we have a continuous regime of rearrangement of the dynamics of multifractality in the trapezius, in the abdominals and the adductors during the pre-NPT and post-NPT treatment.

We deduce that the NPT treatment, modifying the multifractal regime, induce in all the trapezius, abdominals and adductors, modifications that are reflected also in the pre-, during- and post-treatment in the different values of the coefficients A, B, C, that reflect the multifractal spectrum $f(\alpha)$.$\Delta f(\alpha)$ is the difference between the maximum and minimum values of $f(\alpha)$. The difference $\Delta f(\alpha)$ between maximum and minimum values of the singularity provides an estimate of the spread in changes in fractal patterns. Since $\Delta f(\alpha)$ denotes the frequency ratio of the largest to the smallest fluctuation, $\Delta f(\alpha) > 0$ means that the largest fluctuations are more frequent than smallest fluctuations, while $\Delta f(\alpha) < 0$ is the reverse. It is evident from the data of Table 1 that $\Delta f(\alpha)$ are all negative for right trapezius, and for the abdominals, both right and left, while it is positive for the right adductors. For the left trapezius it starts to be positive at rest, pre-NPT treatment, and becomes negative in the NPT treatment and in post-NPT treatment. The same thing happens for the right abdominals. This is the situation of the rearrangements of the multifractal spectrum following the transitions that are induced from the NPT treatment.

In the statistical analysis, the results are significant for h(2) in the left trapezius, for $\Delta\alpha$ in the left trapezius and the left abdominals, for Δf in the right trapezius, Δh in the left trapezius and the left abdominals and for $\Delta\alpha_L$ and $\Delta\alpha_R$ in the left trapezius and left abdominals. The coefficients A, B and C, are statistically significant for the left trapezius and left abdominals. The level of significance is reported with stars in Table 2.

Conclusion

Surface electromyography (sEMG), the nonlinear summation of the electrical activity of the motor units in a muscle which reflects the state of neuromuscular function, has been widely applied to disease diagnosis, pathologic analysis, and rehabilitation evaluation [30–32]. Traditional linear and statistical analysis methods have some significant limitations due to the short-term stationary and lower signal-noise ratio of sEMG, but it is mixed to the nonlinear regime that gives us the main feature of the signal [33].

The basic feature of the linear regime may be analyzed by the Fourier Transform (FFT) with estimation of other linear indexes such as the Muscle Response and the Muscle Fatigue indexes. However, sEMG displays chaotic features and muscle fatigue results in reduced complexity [34] . In this paper, we report multifractal analysis into the field of sEMG process to investigate this dynamical regime: the sEMG on the bilateral masseter, levator labii superioris and frontalis in athletes was recorded. Generally speaking, also the chaotic analysis should have been employed to extract new features, including correlation dimension, Lyapunov exponent, and entropy to indicate in an explicit way that, according to our analysis, the sEMG is multifractal, and chaotic. Consequently, chaotic analysis and multifractal analysis could provide a new insight into the study of the complexity of the EMG and the analysis of multifractality and chaoticity of such a signal could help to analyze the system itself, in particular, the dynamics of modifications found between discrete system's states. The multifractal analysis, that is characteristic of the complexity of the curve, can be computed directly in the time domain by using some algorithms specifically developed. Other studies have shown that the fractal dimension of EMG is sensitive to magnitude and rate of muscle force generated [35–37] and the fractal dimension correlates to the muscle's potential in athletics. The better the athlete is, the higher the fractal dimension. The fractal dimension is representative of the training level. The purpose of this work was to study, in terms of multifractal analysis and to show how the sEMG signals are generated in trapezius, adductors and abdominals pre-, during- and post-NeuroPhysics Treatment providing experimental support to the thesis elaborated in Ross and Ware's previous work [1]. From a clinician and medical perspective, a net modification in the dynamics of the recorded signals may be observed during the treatment. As illustrated by Table 1, we have executed the multifractal analysis of the recorded signals and derived the indices of the Generalized Hurst Exponent, the Generalized Multifractal Dimension, the Fractal Spectrum, and the indices Δh, $\Delta\alpha$, $\Delta\alpha_{\Lambda}$, $\Delta\alpha_R$, $f(\alpha)$, $\Delta\phi(\alpha)$.

All such indices have enabled us to conclude that the sEMG signals are differentiated at rest and during NPT treatment but remain as multifractals. Often the calculated indices show a statistical significance difference between at rest and NPT treatment or in the (left and right) trapezius and (left and right) adductors or between adductors (left or right) and abdominals (left and right).

All these results have been reported in the table of statistical analysis for which we have employed the method of Anova. The scheme of the arising model is as follows: all the muscles, trapezius, abdominal, and adductors represent complex systems whose bio signal profiles, recorded by the sEMG, constitute a multifractal.

These systems are subjected to continuous transitions during the NPT treatment which enable a continuous and differential rearrangement of the indices between different muscles or between the same muscles during the treatment. Such transitions are finalized to maintain such multifractal regime in the muscles. The multifractality, therefore, remains the principal objective of the muscle dynamics that rearranges itself in the occasion of the treatment to maintain such behavior during the phases of pre-, during- and post-treatment. Therefore, we conclude that the exact dynamics of the trapezius, abdominals and adductors is that one of fractality that by continuous transitions, rearrange continuously their dynamics. The attractors of such systems are continuously moving in their inner structure to enable modifications of their structure by transitions that we see by a modification of the multifractal regime.

In conclusion, when performed across all patients of interest, multifractal analysis enables consistency to be accomplished and established. Multifractal analysis of electrophysical data harvested from all patients of interest on day

1 pre-NPT can be performed as a baseline measurement and can then be compared against the multifractal analysis performed on the EMG data from the healthy elite athletes portrayed in this submission.

Author Contributions

Conceptualization, F.C., F.S., R.M., K.W. and E.C.; methodology, F.C., F.S., R.M., K.W. and E.C.; Software, F.C., F.S., R.M., K.W. and E.C.; Validation, F.C., F.S., R.M., K.W. and E.C.; Formal analysis, C.R., F.K., F.C., F.S., R.M., K.W. E.C. and A.D.L.; Investigation, C.R., F.C., F.S., R.M., K.W., E.C. and A.D.L.; Resources, K.W. and E.C.; Data Curation, C.R., F.K.; F.C., F.S., R.M., K.W. and E.C.; Writing—Review & Editing, C.R., F.K., F.C., F.S., R.M., K.W., E.C. and A.D.L.; Visualization, C.R., F.K., F.C., F.S., R.M., K.W., E.C. and A.D.L.; Supervision, C.R., F.K., F.C., F.S., R.M., K.W., E.C. and A.D.L.; Project Administration, C.R., F.K., F.C., F.S., R.M., K.W., E.C. and A.D.L. All authors have read and approved the final version of the manuscript.

Acknowledgments

The authors declare they have no conflict of interest with this work. This research received no external funding. Informed consent was obtained from all subjects involved in the study. The participants (trained athletes) were clients of the NeuroPhysics Therapy Clinic. Collecting electrophysiological data from all clients/patients was approved as a standard procedure by the Association for Practising Neurotricians and Neurotricionists (APNN) (protocol code APNN-2013-0214 and date of approval 10-Feb-2014). Data is available on request. Pre-, during- and post-NPT data from various sources are stored, inclusive of fine-grained gait and stance analysis using Zebris technology at the Institute Neurotricional Sciences, Australia. AusIndustry-Science and Technology Department audits the Institute Neurotricional Sciences every year and data management and patient privacy are also covered in the audit.

References

[1] Namazi, H. (2021). Complexity-Based Analysis of the Correlation between Stride Interval Variability and Muscle Reaction at Different Walking Speeds. *Biomed. Signal Process. Control*, 69, 102956.

[2] Namazi, H. (2019). Decoding of Hand Gestures by Fractal Analysis of Electromyography (Emg) Signal. *Fractals*, 27, 1950022.

[3] Beretta Piccoli, M., Boccia, G., Ponti, T., Clijsen, R., Barbero, M., and Cescon, C. (2018). Relationship between Isometric Muscle Force and Fractal Dimension of Surface Electromyogram. *Biomed Res Int*, 2018, 5373846.

[4] Ross, S. N., and Ware, K. (2013). Hypothesizing the Body's Genius to Trigger and Self-Organize Its Healing: 25 Years Using a Standardized Neurophysics Therapy. *Front. Physiol.*, 4 NOV, doi:10.3389/fphys.2013.00334.

[5] Morrison, S., and Sosnoff, J. J. (2010). The Impact of Localized Fatigue on Contralateral Tremor and Muscle Activity Is Exacerbated by Standing Posture. *Electromyogr Kinesiol*, 20, 1211–1218.

[6] Fink, P. W., Kelso, S., Jirsa, V. K., and De Guzman, G. (2000). Recruitment of Degrees of Freedom Stabilizes Coordination. *J. Exp. Psychol. Hum. Percept. Perform.*, 26, 671–692, doi:10.1037/0096-1523.26.2.671.

[7] Fuchs, A., and Jirsa, V. K. (2000). The HKB Model Revisited: How Varying the Degree of Symmetry Controls Dynamics. *Hum Mov Sci*, 19, 425–449.

[8] West, B. (2013). *Fractal Physiology and Chaos in Medicine*; 2nd ed.; World Scientific, Singapore.

[9] Hong, S., Bodfish, J., and Newell, K. (2006). Power Law-Scaling for Macroscopic Entropy and Microscopic Complexity: Evidence from Human Movement and Posture. *Chaos*, 16, 013135.

[10] Rogers, B., Gronwald, T., and Mourot, L. (2021). Analysis of Fractal Correlation Properties of Heart Rate Variability during an Initial Session of Eccentric Cycling. *Int J Env. Res Public Heal.*, 18, 10426.

[11] Lipsitz, L. A. (2008). Dynamic Models for the Study of Frailty. *Mech Ageing Dev*, 128, 675–676.

[12] Seely, A. J. E., and Macklem, P. (2012). Fractal Variability: An Emergent Property of Complex Dissipative Systems. *Chaos*, 22, 013108.

[13] Heffernan, K. S., Sosnoff, J. J., Jae, S. Y., Gates, G. J., and Fernhall, B. (2008). Acute Resistance Exercise Reduces Heart Rate Complexity and Increases QTc Interval. *Int. J. Sports Med.*, 29, 289–293, doi:10.1055/s-2007-965363.

[14] Izquierdo, I., Bevilaqua, L. R. M., Rossato, J. I., Bonini, J. S., Medina, J. H., and Cammarota, M. (2006). Different Molecular Cascades in Different Sites of the Brain Control Memory Consolidation. *Trends Neurosci.*, 29, 496–505, doi:10.1016/j.tins.2006.07.005.

[15] Burggren, W. W., and Monticino, M. G. (2005). Assessing Physiological Complexity. *J. Exp. Biol.*, 208, 3221–3232, doi:10.1242/JEB.01762.

[16] Kelso, J. A. S. (1995). *Dynamic Patterns: The Self-Organization of Brain and Behavior*; MIT Press, Cambridge, MA, U.S.A.

[17] Iberall, A. S., and Soodak, H. (1998). *Primer on Homeokinetics: A Physical Foundation for Complex SystemsNo Title*; Cri-de-Coeur Press, Laguna Woods, California.

[18] Tsuda, I. (2009). Hypotheses on the Functional Roles of Chaotic Transitory Dynamics. *Chaos*, 19, doi:10.1063/1.3076393.

[19] Kaneko, K., and Tsuda, I. (2003). Chaotic Itinerancy. *Chaos*, 13, 926–936.

[20] Thelen, E., and Smith, L. B. (1994). *A Dynamic Systems Approach to the Development of Cognition and Action*; MIT Press, Cambridge, Massachusetts.

[21] Matthews, P. C., R. E., M., and Strogatz, S. H. (1991). Dynamics of a Large System of Coupled Nonlinear Oscillators. *Phys. D Nonlinear Phenom.*, 52.

[22] Cross, M. C., and Hohenberg, P. C. (1993). Pattern Formation Outside of Equilibrium. *Rev Mod Phys*, 65, 851–1112.

[23] Buchanan, J. J. (1993). Posturally Induced Transitions in Rhythmic Multi Joint Limb Movements. *Exp Brain Res*, 94, 131–142.

[24] Marri, K., and Swaminathan, R. (2015). Analyzing Origin of Multifractality of Surface Electromyography Signals in Dynamic Contractions. *J. Nanotechnol. Eng. Med.*, 6, 031002.

[25] Trybek, P., Nowakowski, M., and Machura, L. (2018). Multifractal Characteristics of External Anal Sphincter Based on SEMG Signals. *Med. Eng. Phys.*, 55, 9–15, doi:10.1016/j.medengphy.2018.03.007.

[26] Wang, F., Yang, Z., and Wang, L. (2016). Detecting and Quantifying Cross-Correlations by Analogous Multifractal Height Cross-Correlation Analysis. *Phys. A Stat. Mech. its Appl.*, 444(C), 954–962.

[27] Agterberg, F. P. (2001). Multifractal Simulation of Geochemical Map Patterns. In *Geologic modeling and simulation: sedimentary systems*; Merriam, D.F., Davis, J.C., Eds.; Kluwer Academic & Plenum Publishers: New York, pp. 327–346.

[28] Eversetz, C. J. G., and Mandelbrot, B. B. (1992). Multifractal Measures. In *Chaos and Fractals*; Peitgen, H.O., Jurgens, H., Saupe, D., Eds.; Springer: New York, pp. 922–953.

[29] Xie, S., and Bao, Z. (2004). Fractal and Multifractal Properties of Geochemical Fields. *Math. Geol. 367*2004, 36, 847–864, doi:10.1023/B:MATG.0000041182.70233.47.

[30] Marri, K., and Swaminathan, R. (2016). Classification of Muscle Fatigue Using Surface Electromyography Signals and Multifractals. In *Proceedings of the 2015 12th International Conference on Fuzzy Systems and Knowledge Discovery*, FSKD 2015; Institute of Electrical and Electronics Engineers Inc., January 13; pp. 669–674.

[31] Marri, K., and Swaminathan, R. (2015). Analysis of Biceps Brachii Muscles in Dynamic Contraction Using SEMG Signals and Multifractal DMA Algorithm. *Int. J. Signal Process. Syst.*, 4, doi:10.12720/ijsps.4.1.79-85.

[32] Chatterjee, S., Roy, S. S., Bose, R., and Pratiher, S. (2020). Feature Extraction from Multifractal Spectrum of Electromyograms for Diagnosis of Neuromuscular Disorders. *IET Sci. Meas. Technol.*, 14, 817–824, doi:10.1049/iet-smt.2019.0132.

[33] Arjunan, S. P., and Kumar, D. (2007). Fractal Theory Based Non-Linear Analysis of SEMG.; *Proc. 3rd International Conference on Intelligent Sensors, Sensor Networks and Information*, pp. 545–548.

[34] Rampichini, S., Vieira, T. M., Castiglioni, P., and Merati, G. (2020). Complexity Analysis of Surface Electromyography for Assessing the Myoelectric Manifestation of Muscle Fatigue: A Review. *Entropy*, 22, doi:10.3390/E22050529.

[35] Wang, G., Ren, X. M., Li, L., and Wang, Z. Z. (2007). Multifractal Analysis of Surface EMG Signals for Assessing Muscle Fatigue during Static Contractions. *J. Zhejiang Univ. Sci. A*, 8, 910–915, doi: 10.1631/jzus.2007.A0910.

[36] Wang, G., and Ren, D. (2013). Classification of Surface Electromyographic Signals by Means of Multifractal Singularity Spectrum. *Med Biol Eng Comput*, 51, 277–284, doi: 10.1007/s11517-012-0990-9.

[37] Marri, K., and Swaminathan, R. (2015). Identification of Onset of Fatigue in Biceps Brachii Muscles Using Surface EMG and Multifractal DMA Algorithm. In *Proceedings of the 52nd Annual Rocky Mountain Bioengineering Symposium and 52nd International ISA Biomedical Sciences Instrumentation Symposium 2015*, pp. 109–116.

Chapter 6

Integrated Psychosomatic-Somatopsychic Training in a Systemic Approach for Youth Competitive Sport Technicians*

Laura Cancellara*
and Sonia Ballocco
PM Mind Force Society/MIUR Roma, Italy

Abstract

For technical training it is necessary to consider as peculiar the training that integrates a psychosomatic-somatopsychic approach for understanding the stages of development of the agonistic Self. In order for the athlete to reach global awareness of the Self, it is important that the technical team has adequate training that integrates an approach to bodily, emotional, cognitive, relational and competitive awareness of the Self. In this way the athlete will be able to master his skills, self-regulating in moments of emotional imbalance, in moments of cognitive confusion and body dissociation reaching an optimal state of flow during performances.

Keywords: emotions, global self awareness, body awareness, emotional awareness, competitive self, psychosomatic-somatopsychic blocks, PMP, psychosomatic mindfulness protocol

---

* This is a reformatted version of an article previously published in *Chaos and Complexity Letters*, Volume 16 Issue 3 (2022), New York: Nova Science Publishers.

* Corresponding Author's E-mail: laura.cancellara@gmail.com.

In: Complexity Science in Human Change
Editors: Franco F. Orsucci and Nicoletta Sala
ISBN: 979-8-89113-408-9

Introduction

The proposal in question, points the focus of attention on the importance of including, in the training of the technical-sports programs of the coaches, the issues related to the bodily self-awareness of the competitive athlete. Different international reviews that pay attention to high-level athletes bring out worrying news about the psychophysical well-being of young athletes who, in pre-adolescent and adolescent age, are subjected to very intense physical training and which often compromise the quality of the sporting lived experience, causing an early sports drop-out. In most of the sports practiced during adolescence, intensive and continuous training exposes young athletes to various stresses (psychic, physical and relational) already in the pre-pubertal age. If we think of the aesthetic disciplines (professional dance, artistic gymnastics, and rhythmic gymnastics), have their maximum performance during the adolescence phase, due to the proportions of the body.

Attention to physical performance alone, to the achievement of optimal performance, often limits the collection of experience in the emotional and socio-relational areas. The lack of attention to the dimensions of a balanced work life balance could become a factor of vulnerability and pathological distress, especially in crisis situations or when their athlete's career suddenly ends. Moreover, especially when athletes are oriented towards achieving great sporting successes, they are exposed to high social expectations. It is not uncommon that in these situations a dysfunctional condition related to sporting life can occur which causes burnout, sports dropout up to the possibility of a psychic or psychosomatic pathology. Equally stressful can be the reduction of training skills in adulthood and the realization that the pinnacle of a career has been reached in the first third of life. Incisive attitude perfectionism, scarce knowself-esteem, emotional dysregulation, deficit of coping strategies for stress management, obsessive traits, seem to be personality characteristics that are found with good frequency in high-level athletes. Situations of protracted stress can intervene, allowing suffering to emerge psychophysics. They come so, to be urged, not only physical and athletic abilities but also mental ones and, specifically, metacognitive ones. The presence of malfunctions of capabilities metacognitive, emphasized by the constant activation of the agonistic system, is able to produce discomfort and psychic suffering and reduce the ability to solve problems and therefore, as such, a privileged object of attention in training programs for technical managers. In this article, we discuss the key components and potential brain-body mechanisms related to psychological well-being and propose global awareness training as a systemic and integrated approach.

The purpose of this chapter is to provide useful material to help technical sports professionals increase awareness both of their own thoughts, feelings, desires, and behaviors, and to ensure that the needs of the athletes they train are met. The operationalization of educational and training strategies based on the acquisition of transversal skills, (as for students training) considers data relating to the personality traits, personal goals, motivations and personological characteristics: conscientiousness, intrinsic goal orientation, metacognitive self-regulation (Cancellara, et al., 2013). This may mean that to improve the quality of sports performance on the psycho-educational side, the coach becomes "sensemaker."

It is proposed in this regard to:

1. *To promote* the acquisition of the necessary and most up-to-date skills sports technicians for favor complex and integrated reasoning of the mind-body relationship and the overall awareness of the athlete's self.
2. *Implement* methodologies, techniques, and strategies to foster a meaningful empathic relationship with the young man athlete which pays attention to emotional experiences and bodily sensations.
3. *To distribute* training loads adequately with reference to the characteristics of adolescence, in order to prevent psychosomatic disorders, distress and early abandonment.
4. *Monitor and evaluate* the correlated psycho-educational variables and promote career longevity.
5. *Consider* from a salute genic and integrated point of view, the processes of psychic growth and empowerment of the athlete to promote health, in order to better express conscious choices, aimed at psychophysical well-being.

The proposed intervention will also serve to make sports technicians aware of problems relating, for example, to the management of competitive stress and anxiety, overtraining syndromes (overtraining, overreaching) or for the optimization of attention skills or relational difficulties (relating to the relationship between teammates or between coach and athlete). Sometimes, young athletes, perceive the relationship between mind and body as not integrated or to their own emotional states and personal experiences. The link between emotional states and stressful events they take a central aspect, nevertheless, often young athletes, they do not grasp the relationship between emotional experiences and physical states. The two experiences are commonly unrelated, and it is precisely this split that does not facilitate the recovery of a state of psycho-physical well-being during and after an intense workout.

Combining like this, the narration of the athlete with the observation of his body dimension, we are able to arrive at a more correct reading of his experiences, favoring an increase in one's self- awareness, of one's own psychosomatic patterns of instinctive, emotional and psychological behavior which will be useful for directing the effectiveness of sports training.

Theoretical Background

A growing literature supports the idea that there is an important relationship between psychological well-being and mental and physical health in both adolescents and adults (Ong et al., 2011; DeSteno et al., 2013). Psychological well-being is a multidimensional construct that pertains to optimal psychological functioning and experience.

Preventing the psychophysical health of the athlete promotes resilience understood as a coping strategy, self-regulation of emotions, oriented problem solving. To promote holistic wellness, it is helpful to understand the underlying mechanisms associated with this construct in order to develop effective training programs. We focus on one of the mechanisms through which IBMT (mind body integration) improving psychological well-being. In fact, the interaction between mind (awareness) and body (bodyfulness) - involves both the central nervous system (CNS) and the autonomic nervous system (ANS).

In a series of randomized controlled trials (RCTs), IBMT (Integrative Body-Mind Training) has shown multiple positive effects on psychological well-being and health, including increased self- control and positive emotions, and decreased negative emotions and stress hormones. However, it is necessary to know that although these results are promising (Chambers et al., 2009; Hölzel et al., 2011), neurobiological studies for mindfulness training practices directly correlate both behavioral changes and functional brain activity are still scarce (Fox et al., 2016).

We can suggest that mindfulness training may be a promising program that promote the synergistic engagement of mind, body, and context, to achieve the goals of enhancing psychological well-being during training sessions (Tang et al., 2015; Ding et al., 2014).

Having a greater global self-awareness allows us to consider what are the preventive factors for many health problems in youth sport (Kerut, 2014). For example, eating disorders can be present in wrestling and in aesthetic sports such as professional dance, artistic gymnastics, and rhythmic gymnastics. In fact, placing too much emphasis on quantitative (body size, weight, height) and qualitative aspects such as the focus on body image as prerequisites for sporting success aggravates the mental health condition of the young athlete. For rhythmic gymnastics athletes for example, "success is strongly influenced by visual appeal and body aesthetics". Rhythmic gymnasts are often required to meet certain weight targets to attain and maintain a thin shape (Klinowski et al., 2007). The pressure for personal appearance and the control of qualitative indicators (to the detriment of the athlete's global well-being) is intense and, consequently, deficits can be diagnosed from: body dysperception, eating disorders, delayed menarche, menstrual irregularities, low body fat, and delayed maturity (Klinowski et al., 2007). We therefore speak of dysfunctional behaviors that distance the athlete from his own psychophysical and emotional well-being and from a perception of complete self-awareness.

Emotional well-being has been defined as an overall positive state of one's emotions, life satisfaction, sense of meaning and purpose, and ability to pursue self-defined goals.

Elements of emotional well-being include a sense of balance in emotions, thoughts, social relationships, and activities. Several systematic reviews and meta-analyses have suggested that psychological well-being can be improved by interventions such as mindfulness training (Kong et al., 2016; Feller et al., 2018; McConville et al., 2017).

As Burt Giges (Past-President of the Association for Applied Sport Psychology) suggested, conceptualizing the origins of thoughts, feelings, and behaviors may be a necessary skill that training specialists (including sports psychologists) can develop to better serve their clients and foster self-awareness (Giges, Reid, 2016).

The approach to considering Psychosomatic Mindfulness Protocols, the contributions of third-wave cognitive-behavioral therapies, and attention to somatic competence can guide technicians to a profound awareness of several dimensions. Thoughts (e.g., beliefs limiting, mechanical thoughts), feelings (regulation of emotions and somatic sensations) and control of performance (adaptive or maladaptive) are important dimensions that need to be integrated with each other in a harmonious and balanced way.

Integrated Somatic Competence

In recent years, psychosomatic medicine has had renewed interest from the international scientific world, so much so that it was considered in 2003 as the seventh psychiatric ultra-specialization by the American Board of Medical Specialties in the USA[2]. With a holistic and integrated vision (from the Greek "olos" which means "whole," conveys a sense of totality) which focuses on the relationship between the subject and its complexity, the contribution of Psychosomatic Psychology (and therefore, of a sports psychotherapist with such specialization requirement) will help to extend the possibilities of listening and intervention with the characteristic of a team approach (comparison between sports technicians, clinicians, and athletes). The model of Psychosomatic Psychology intervenes according to a systemic perspective of integration between body and mind to re-establish a relationship of balance between the person and his suffering (such as alterations of the emotional aspect with possible pathogenetic significance).

The sports coach, to ensure attention to a state of integration, will have to consider the awareness that mind-body-relationships (biopsychosocial model, originally advanced by George L. Engel, 1977) are interrelated in a dynamic continuum of adaptation to the challenges of the external environment. Sometimes, the young athlete fails to integrate emotional experience and psychophysical states. Distress from overtraining and related psychosomatic disorders, require a complex and multifactorial reading: the consequences deriving from somatization can in fact make the athlete vulnerable in many forms. The same amplification of bodily sensations, that is the tendency to experience somatic sensations as intense or dysfunctional, implies psychological processes of hypervigilance or increased attention to bodily sensations, the tendency to exasperate weak and infrequent sensations, a marked reactivity to bodily sensations with emotions and thoughts that intensify them, making them more disturbing, alarming, and ominous (Barsky et al., 1988; Pennebaker & Skelton, 1981).

Also considering other dimensions such as nutrition, musculoskeletal problems, interpersonal relationships, means systemically considering variables that influence the psycho-biological balance of the young athlete. For this reason, knowledge of the psychosomatic dimension will serve to potentially identify both the athlete's strengths and psycho-physical breaking

---

[2] ISTUD- Mind-body/body-mind dialogue: health outcomes and prospects for sustainability- Program Scientists in the Company - XII Edition Stresa, 26 September 2011 - 27 July 2012.

points, in order to try to re-establish recovery connections even within the somato-psychic network.

Paying attention to a dimension that integrates psychosomatic and somatopsychic networks will lead to a greater global awareness of the athlete's self. This means implementing a real training program that pays attention to "sports care", including both the somato-psychic and systemic components (belonging context and family history). Paying attention to these dimensions also means being able to understand internal and externalized "signals", learning to distinguish, for example, between organic pain and emotional experiences. For the sports technician, it is a question of integrating the professional technical skills of 'knowledge' and 'know-how' with the skills of emotional-relational intelligence and "knowing how to be". This capacity is essential for recognition of feelings which, consequently, it allows awareness and tuning on the athlete's experiences, on the relational alliance and on the construction of a more solid competitive self. Training Somatic Competence means listening to one's body, observing the sensations that emerge and expressing them and modifying one's internal states to gradually regain awareness, orientation, and balance. The increase in Somatic Competence can therefore help prevent dysregulated behaviors, more generally, improve the quality of relationships.

Psychosomatic Psychology therefore promotes a culture of "being with", whether it is personal experiences, the external context, or the relationship with the other, for the dynamic and conscious development of training objectives.

In this regard, it is essential to consider the athlete an active ally in the training process (characteristic of agency), so the athlete's goal will not only be to train for optimal performance, but to be able to build together with the coach, the ability to take care of yourself. Having an integrated somatic competence, therefore, for the sports coach and for the young athlete means to be more aware of the dimension of the competitive self: learning to be in tune with one's body, understanding and distinguishing its emotional signals, knowing how to recover from states of imbalance inside during sports training, and more generally, taking care of one's mental and physical health tout court.

The Role of Genetic and Epigenetic Factors for the Construction of the Agonistic- Self

Before becoming an athlete, says Allan Schore, the child is born with his own self: the implicit self. The existence of this first Self is unconscious and is expressed through the body with non-verbal language. All the development of the individual system and of the global self which includes the formation of the competitive self of the athlete is determined by the quality of the primary relationships, as a first imprinting in the development of the structural systems of the right brain involved in the unconscious elaboration of emotion, in the modulation of stress, in self-regulation and therefore in the functional origins of the affective center of the body-based implicit self (Ghiroldi, 2018). With simple contact with the baby's skin, the mother acts as a regulator of the baby's internal homeostasis.

Furthermore, according to Cloninger's theory, the temperament from which the differences between individuals are identified in their automatic responses to stimuli can be identified in four dimensions to which the biological and genetic correlates have been analysed:

- *Search for novelty* (dopaminergic tone): people in this group have high levels of stimulation due to an excessive production of dopamine, which they manifest with a tendency towards exploration, enthusiasm, impulsivity, relational instability and disorder, tendency to boredom, fickleness in decision-making and sudden outbursts of temper.
- *Damage avoidance* (hyposerotonergic tone): this group instead includes people with a high sensitivity to criticism and punishment, are concerned about the consequences of their actions and are characterized by low energy levels and a tendency to be chronically exhausted.
- *Reward addiction* (hyponoradrenergic tone): people classified in this tone have an important social vulnerability, they tend to be very shy up to social phobia and social anxiety, worrying about other people's reactions to their behavior.
- *Persistence* Finally, in this group, which represents the fourth dimension of Cloninger's temperament, we find people with the ability to persevere with determination and constancy. They are the most difficult to extinguish inappropriate behavior, on the other hand they are the most ambitious and perfectionist.

Each dimension describes the characteristics of the temperament from which to start in the evaluation and help of the athlete and of the individual in general. Know e.g., if faced with an inner discomfort, the athlete tends to stretch out or stiffen, it is important for the formulation of a personalized competitive athletic training.

Each of these four temperament dimensions is approximately 50% heritable (Gillespie NA, 2000). The genetic component is 50% important compared to the epigenetic one which, as the evolutionary stages go through, acquires less weight in adaptability to the environment.

A systemic approach that integrates physical proprioception with other levels of the Self will allow us to intervene in the clinical process regarding rigidity and automatisms such as the reiteration of repeated responses to specific external stimuli which are automatic and unconscious.

Sphere of the Psychosomatic Self: Psychosomatic Disconnection and Integrated Awareness

Jaak Panksepp, psychobiologist and neuroscience researcher at Washington State University, using the ESB, Electrical brain stimulation, highlighted the existence of seven main emotional systems:

1. the FEAR/ANXIETY system, linked to cortisol.
2. the SEARCH/ENTHUSIASM system, linked to dopamine.
3. the system of SEXUALITY/DESIRE and lust, linked to sex hormones.
4. the ANGER/DOMINANCE system, linked to testosterone and serotonin.
5. the CARE/LOVINGNESS system, linked to oxytocin.
6. the system of SADNESS/PANIC, linked to the absence of CARE.
7. the system of PLAY/FANTASY and socialization, connected to the hormones of dopamine and endorphin.

According to Panksepp in real life, the Self manifests itself through these seven primary emotional systems and their alteration or inhibition is at the origin of the main psychosomatic illnesses and personality disorders.

Individual differences are due to the different proportions and expressions of these seven primary elements which form the basis for higher order complex emotions: pride, shame, jealousy, guilt, security, etc.

Each emotional system represents a "psychosomatic functional structure", essential to life, which integrates and manifests itself simultaneously with psychological-neocortical, emotional-limbic, and behavioral-instinctive aspects, as well as a "PNEI system". Each individual has all the circuits and, potentially, all possible gradations of emotional tonality and psychosomatic structure (Figure 1). Nine major neurotransmitters have been identified that are present at the level of psychological, somatic, and emotional activities: serotonin, dopamine, testosterone, cortisol, adrenaline, norepinephrine, oxytocin, vasopressin, and endorphin.

The ultimate goal is to identify the neuropersonality (Figure 2), i.e., the "emotional based personality," through scales and tests, but above all through a careful analysis and evaluation of the behavioral, emotional and psychological psychosomatic characteristics that identify the person's temperament. This allows to become aware of one's own psychosomatic patterns of instinctive, emotional, and psychological behavior and in particular, for doctors, psychologists, psychiatrists and sports psychotherapists and others, to possess specific knowledge on how to rebalance the inhibited neuropersonalities or hyperactive and bring the person back to a state of balance and psychophysical health. With this fundamental stage, all the targeted training schedules and programs will be personalized and more effective.

The Affective Neuroscience Personality Scale (ANPS) is a psychological test, developed starting from Panksepp's research, to evaluate the characteristic behavioral traits of the seven emotional systems and, indirectly, also the different neuropersonalities.

Every alteration, inhibition or hyperactivation of a primary or secondary emotional system affects the overall balance and therefore the well-being of the entire system. Edelman (Cloninger 1999) and Giulio Tononi, identify in the thalamocortical circuit the dynamic network of consciousness, the central cerebral network that regulates the information of the entire neuropsychosomatic network, the neuronal basis of consciousness (Edelman G., 1998).

Figure 1. Panksepp, the seven emotional system.

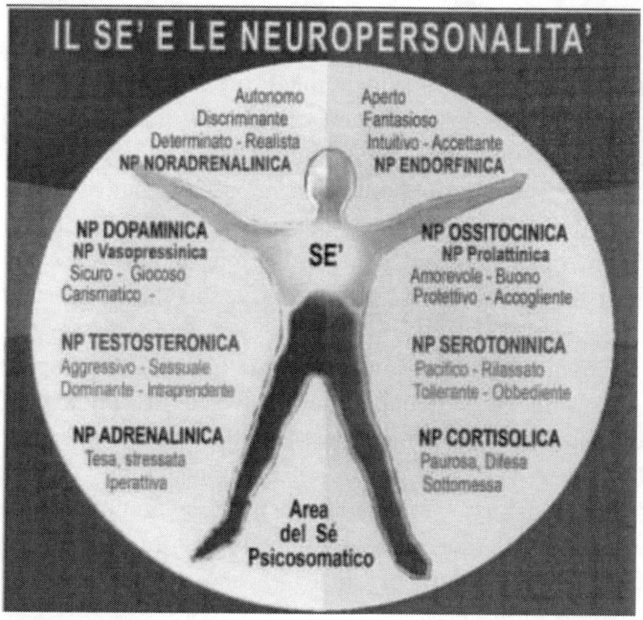

Figure 2. Montecucco, the self and neuropersonalities.

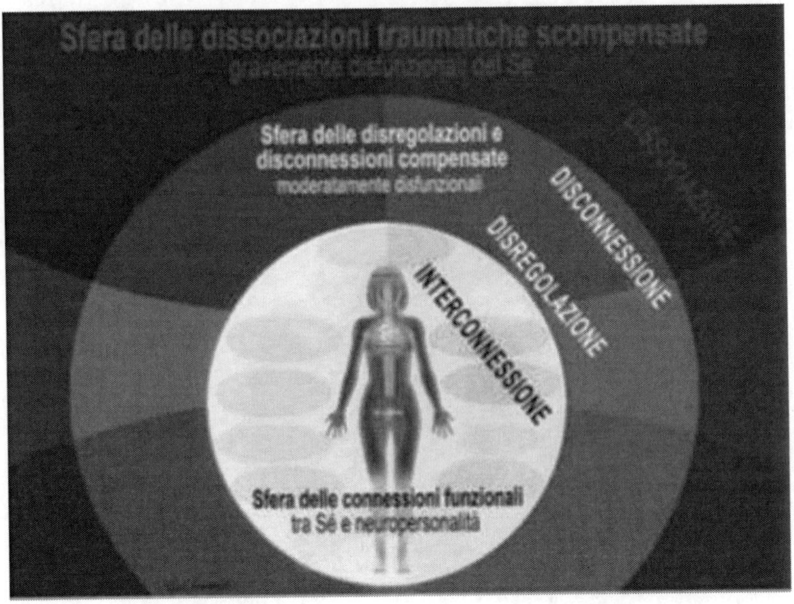

Figure 3. Montecucco, sphere of decompensated traumatic dissociations.

The evidence that the awareness of the Self is strongly centered on the thalamus is also confirmed by the fact that the damage to the thalamus corresponds to a state of permanent coma. From years of clinical observations, FN Montecucco shares his profound conviction that the thalamic function reflects, at a somatic level, the cardiac function. The thalamus allows information to circulate throughout the brain just as the heart allows blood and hormones to circulate throughout the body. For a correct diagnostic evaluation and for effective psychosomatic therapy before and during athletic training, understanding the genesis and structure of psychosomatic disorders and blocks is a central element.

Often the progressive establishment of the "neuropsycho-somatic disconnection" hinders the integration between the parts of the system-network and, initially, generates processes of functional dysregulation of the emotional systems (first cluster), then phenomena of "fragmentation" or psychosomatic "blockage" (second cluster), to the point of degenerating in the most serious cases into traumatic "dissociation" (third cluster). Thus, the term block refers to any form of neuro-psycho- somatic dysregulation or disconnection (Figure 3).

From a gene-environment interaction perspective, or from a biopsychosocial model, some cases are considered in which the primary cause of psychosomatic disorders may derive from a genetic or traumatic alteration that modifies the intertwined communications between the emotional systems and PNEI (subsystems of the psychosomatic unit), in other cases that the cause depends on one or more environmental, affective or social events that generate imbalance in the entire system.

Clinical Framework of Psychosomatic Disorders

In psychosomatic disorders we can recognize three large clusters of psychosomatic disorders that require different clinical approaches: non-specific general disorders, emotional disorders, and self- disorders. As regards the general blockages (psychosomatic disorders) caused by stress, we recognize them in the forms of alteration of response to stressors: illnesses, accidents, excess work, difficult economic commitments (debts, mortgages), significant life changes (change of house, relocation, retirement), physical or intellectual effort, excessive natural events (heat, cold, earthquake and similar catastrophes). These progressive tensions manifest themselves on all PNEI systems and are reflected in a cascade by altering all the main physiological systems of the neuropsychosomatic network with classic symptoms such as: depression, nerve and muscle tension, agitation, anxiety, insomnia, inflammation, joint pain, etc.

These reactions can be objectively evaluated, in a medical or psychological context, especially through six main physiological systems: nervous, muscular, respiratory, gastrointestinal, cardiocirculatory and postural.

Potentially each organ can become a target of stress pathologies. The causes that lead an organ to be the "target" of a psychosomatic disorder are strictly physiological and functional within a PNEI approach, and, except in rare cases, are not linked to unconscious symbolic activities. In psychotherapy, on the other hand, it is possible to recall the original experiences of the disorder through an associative work of a symbolic type. When you experience a situation of danger, fear or pain, to protect yourself, the systems of pleasure are inhibited and all energies are shifted to activate the system of FEAR, the main system of "defence of the Self."

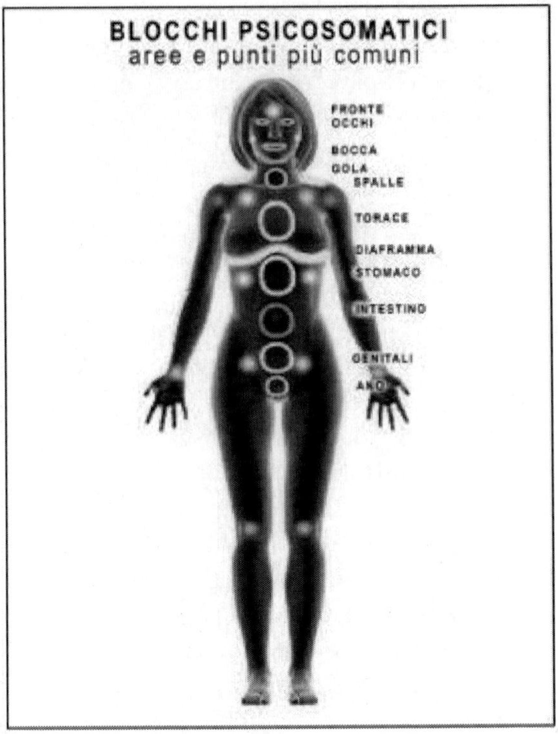

Figure 4. Montecucco Psychosomatic Blocks.

Through the "stress axis" the production of adrenaline, noradrenaline and cortisol is stimulated, allowing to release and use all the available energy of the system (glucose) to support neuro-muscular tension and overcome the difficulty.

The dysfunctional activation of the stress axis develops in four phases:

- functional alarm phase (eurostress),
- resistance phase,
- Exhaustion stage
- phase of dissociation.

Many psychosomatic blocks (Figure 4) derive from the inhibition or hyperactivation of an emotion through a "specific" psychosomatic response characteristic of one or more neuropersonalities (second cluster).

Most of the symptomatic medical, pharmacological, or natural "palliative" treatments do not obtain the expected results of complete restoration of the original well-being because they are linked to the old paradigm. We need new diagnostic tools that are based on specific therapeutic practices of intervention on the blocks and remove the real causes of the disorders. The practices that start from the body aim to modify the primary causes of stress through a change in its physical-energetic structure and behavioral and emotional response, thus bringing it to a different level of ability to manage emotions through words.

We can find "oral blocks" developed by children who have not been breastfed, "eye blocks" due to highly normative parents or superiors (educators, sports technical staff and mental coaches) who use gaze and visual dominance, the common of the throat" from shyness developed by children who have suffered an excessive inhibition of their spontaneous emotional expression by submission or judgment and the "hand blocks" of anxious people, with cold and tense hands from too much adrenaline (like Raynaud's syndrome), organ block (also called "functional") is often intrinsically linked to an overly activated or inhibited emotional system.

We find the classic blocks of the heart characteristic of all the themes of affective disorder of the CARE/LOVE system. Or blocks of the SEXUAL system for religious reasons or excessive moral control, in which the "target organ", in this case the genitals are affected by vaginitis, impotence problems, etc. Or from the generic block of the sexual function that occurs when it is over-controlled and disturbed with the avoidance of intimate relationships, frigidity, anorgasmia. Or again in the blockage of the noradrenaline MENTAL CONTROL system which has repercussions in the stiffness of the back that appears "like a broomstick" ("straight back syndrome"). Or finally in the lack of spontaneity and BODY PLEASURE (anhedonia) which often characterizes the children of families with very harsh upbringing and regulations for example. those with backgrounds in religious colleges, military academies or overly demanding physically and psychologically training courses. The "profound" traumatic psychosomatic blocks instead (third cluster) manifest themselves with alterations of the person's identity, modifying the central consciousness (conscious and unconscious) which directs and coordinates the functions and communications of the whole person and manifesting themselves as phenomena of disconnection, or worse, of psychosomatic dissociation.

The clinical approach of this type of disorder requires the intervention of a sports psychotherapist with integrated training who has himself worked on the profound awareness of himself and who is therefore able to operate with equal intensity on the other person's self, through empathic resonance work.

The Sense-Maker Sports Coach: Psychosomatic Mindfulness Protocols

The use of Mindfulness ("fullness of consciousness", awareness) and Meditation ("state of no mind", voluntary interruption of mental noise, reduction of excess thought activity) have had significant scientific results. We can say that Consciousness represents a state of mind (Simon, 2007). Furthermore. the meditative practice modifies the "neural structure" and the brain activation structure by modifying the related physical and emotional states (Ghiroldi, 2018).

Many therapeutic models are based on a top-down approach to change through verbal representations beliefs, affects, internal operating models and all that is considered fundamental for a transformation of the Self are explored. Alongside this approach, we believe it is essential to integrate bottom-up intervention methodologies, which therefore start from bodily sensations, from breathing, from body contact. Body contact intensifies the feeling of well-being and "emotional and energetic nourishment". It is interesting to remember how it is generated from the cells of the external embryonic layer (ectoderm) which will also generate the skin. The skin will maintain protective, barrier, sensory and communication functions throughout life (Ghiroldi, 2018). It is crucial that contact takes place when there is admissibility. We can mention one of these techniques that has had and is continuing to have many clinical benefits, the Psychosomatic Mindfulness Protocol in the Gaia Project by FN Montecucco.

Several international studies show that the adults, through awareness techniques, such as body scan techniques, have a better chance of fully concentrating on the task and promoting a positive body image (Todd et al., 2019).

Other clear findings support this argument: mindfulness training can improve students' psychological well-being. The psychological well-being of the student, in fact, improves learning skills, social relationships and academic performance (University of Exeter, 2013). These awareness practices could also bring valuable contributions for adolescent athletes, in fact, potentially, they could favor a change in the personal perception of body composition or eating disorders, sometimes related (Todd et al., 2019).

Like any other skill, mastering awareness takes training.

Through awareness training, athletes can experience metacognitive processes, conscious acceptance and functioning of thought processes by engaging in what is good for them (Hansen, Haberl, 2020). Teaching athletes awareness helps them above all to keep their attention on the present moment, favoring the achievement of maximum performance and well-being (Henriksen, 2022).

Conclusion and Future Trends: Psychosomatic Mindfulness Protocol in Sport

The validation of the Psychosomatic Mindfulness Protocol has obtained significant results on different neurophysiological parameters and on psychological tests.

In the first preliminary neuropsychological research, the electroencephalographic data relating to the analysis of "brain coherence" (EEG coherence) and the main neurophysiological parameters related to stress, respectively Heart Rate Variability (HRV), nerve and muscle functions were related with Electromyography (EMG) and electrodermal response with Galvanic Skin Response (GSR). Together with these parameters, clinical symptoms and well-being were measured with the SRT and SQ tests (Symptom Rating Test and Symptom Questionnaire), anxiety and depression with the Zung Test, emotion recognition with the TAS20 (Toronto Alexithymia Scales), mindfulness skills using the FFMQ (Five Facet Mindfulness Questionnaire), psychological and psychosomatic stress with the MSP (Mesure du Stress psychologique) and the ability to empathize with the BEES (Balanced Emotional Empathy Scale). Finally, by submitting the FFMQ (Five Facet Mindfulness Questionnaire), it was possible to notice a significant improvement in the five parameters of mindfulness: observation, description, awareness, non-judgment, non-reaction.

Technicians with relevant specific training and qualifications for the use of the PMP protocol could improve the internal resources of young athletes. Specifically, unnecessary stress should be avoided, and performance fear rebalanced within the sphere of tolerance. Furthermore, we will work on increasing muscle strength without causing psychosomatic blockages, on the safety of our sporting abilities and potential (competitive self) and on relational ones.

The mind-body unity as an inseparable psychobiological unit of the individual constitutes the presupposition of the psychosomatic-somatopsychic circuit on which to base the strategies of psychological support in the use of the aforementioned integrative tools in the training and monitoring of the sportsman.

References

Barsky J D Goodson, R S Lane, P D Cleary. The amplification of somatic symptoms. Psychosomatic Medicine (1988) Cancellara, Laura; Sala, Nicoletta; Appeddu, Paola; Lai, Alessandro; Orsucci, Franco (2021) Multidisciplinary skills in the school of complexity: educational technologies, human factor, and non - cognitive skills. *Chaos and Complexity Letters*, Nova Science Publishers, Inc Hauppauge Vol. 15, Fasc. 2/3.

Cloninger, C. (1999). *Personality and Psychopathology*. American Psychiatric Pub.

Chambers, R., Gullone, E., and Allen, N. B. (2009). Mindful emotion regulation: an integrative review. *Clin. Psychol. Rev*. 29, 560–572. doi: 10.1016/j.cpr.2009.06.005.

DeSteno, D., Gross, J. J., and Kubzansky, L. (2013). Affective science and health: the importance of emotion and emotion regulation. *Health Psychol*. 32, 474–486. doi: 10.1037/a0030259.

Ding, X., Tang, Y. Y., Tang, R., and Posner, M. I. (2014). Improving creativity performance by short-term meditation. *Behav. Brain Funct*. 10:9. doi: 10.1186/1744-9081-10-9.

Edelman G. M., T. G. (1998). Consciousness and Complexity. *Science* Vol 282, Edizione 5395, pp. 1846.1851.

Feller, S. C., Castillo, E. G., Greenberg, J. M., Abascal, P., Van Horn, R., Wells, K. B. (2018). Emotional well-being and public health: proposal for a model national initiative. *Public Health Rep*. 133, 136–141. doi: 10.1177/0033354918754540.

Fox, K. C., Dixon, M. L., Nijeboer, S., Girn, M., Floman, J. L., Lifshitz, M., Melissa Ellamil, Peter Sedlmeier, Kalina Christoff. (2016). Functional neuroanatomy of meditation: a review and meta-analysis of 78 functional neuroimaging investigations. *Neurosci. Biobehav. Rev*. 65, 208–228. doi: 10.1016/j.neubiorev.2016.03.021.

Ghiroldi, S. (2018). *In Il corpo e la riparazione del Sè* (p. 21). Roma: Alpes Italia.

Giges, Reid (2016): *Mindfulness and Performance*, Cambridge University Press pp. 488 – 512. doi: https://doi.org/10.1017/CBO9781139871310.022.

Gillespie N. A., C. C. (2000). In C. C. Gillespie N. A., The genetic and enviromentl reltionship between Cloninger's dimensions of temperment and chracter. *Personality and individual differenes* 35:1931-1946.

Hansen, J., and Haberl, P. 2020. "Helping athletes be present when performing under pressure," in *Mindfulness and Acceptance in Sport: How to Help Athletes Perform and Thrive Under.*

Pressure, eds K. Henriksen, J. Hansen and C. H. Larsen (New York, NY: Routledge). Henriksen K (2022) The Magic of Mindfulness in Sport. *Front. Young Minds*. 10:683827. doi: 10.3389/frym.2022.683827.

Hölzel, B. K., Lazar, S. K., Gard, T., Schuman-Olivier, Z., Vago, D. R., and Ott, U. (2011). How does mindfulness meditation work? Proposing mechanisms of action from a conceptual and neural perspective. *Perspect. Psychol. Sci.* 6, 537–559. doi: 10.1177/1745691611419671.

ISTUD- Mind-body/body-mind dialogue: health outcomes and prospects for sustainability- Program Scientists in the Company - XII Edition Stresa, 26 September 2011 - 27 July 2012.

Kerut, Edmund (July 2014). "Preventing Heat Illness in youth athletes". *Louisiana's Health & Fitness Magazine*. Archived from the original on July 25, 2014.

Klinkowski, Nora; Korte, Alexander; Pfeiffer, Ernst; Lehmkuhl, Ulrike; Salbach-Andrae, Harriet (10 September 2007). "Psychopathology in elite rhythmic gymnasts and anorexia nervosa patients". *European Child & Adolescent Psychiatry*. 17 (2): 108–113. doi: 10.1007/s00787-007-0643-y. PMID 17846815.

Kong, F., Wang, X., Song, Y., and Liu, J. (2016). Brain regions involved in dispositional mindfulness during resting state and their relation with well-being. *Soc. Neurosci.* 11, 331–343. doi: 10.1080/17470919.2015.1092469.

McConville, J., McAleer, R., and Hahne, A. (2017). Mindfulness training for health profession students-the effect of mindfulness training on psychological well-being, learning and clinical performance of health professional students: a systematic review of randomized and non-randomized controlled trials. *Explore* 13, 26–45. doi: 10.1016/j.explore.2016.10.002.

Montecucco, N. F. (s.d.). *Protocollo Mindfulness Psicosomatica PMP.* Lucca: Associazione "Villaggio Globale."

Ong, A. D., Mroczek, D. K., and Riffin, C. (2011). The health significance of positive emotions in adulthood and later life. *Soc. Personal. Psychol. Compass* 5, 538–551. doi: 10.1111/j.1751-9004.2011. 00370.x.

Panksepp, L. B. (2012). *Archeologia della mente*. Origini . Cortina Editori.

Panksepp, J. (2004). In *Affective Neuroscience: The Foundations of Human*. Oxford University Press.

Pennebaker, J. W., & Skelton, J. A. (1981). Selective monitoring of physical sensations. *Journal of Personality and Social Psychology*, 41, 213-223. Schore, A. N. (2010). I Disturbi del Sé. Roma: Astrolabio.

Simon, V. (2007). Mindfulness y neurobiologia. Rivista de Psicoterapia, 66(67):5-30. Tang, Y. Y., Hölzel, B. K., and Posner, M. I. (2015). The neuroscience of mindfulness meditation. *Nat. Rev. Neurosci.* 16, 213–225. doi: 10.1038/nrn3916.

Todd Jennifer, Jane E. Aspell, David Barron, Viren Swami (2019) Multiple dimensions of interoceptive awareness are associated with facets of body image in British adults, *Body Image,* Volume 29, ISSN 1740-1445, https://doi.org/10.1016/j.bodyim.2019.02.003.

University of Exeter. (2013, June 19). Mindfulness can increase wellbeing and reduce stress in school children. *ScienceDaily*. Retrieved March 27, 2023 from www.sciencedaily.com/releases/2013/06/130619195139.htm.

Chapter 7

Rorschach Ink-Blots Are Multifractals, and a Corresponding Multifractal Structure Is Realized in Humans during Their Perception as Measured by the GSR (Galvanic Skin Response) Neurological and Psycho-Physiological Signal*

Sergio Conte[1]
Leonardo Mendolicchio[1,2]
Ferda Kaleagasioglu[1,3]
Rich Norman[1]
Fang Wang[4]
Francesco Casciaro Losito[1,5]
Anna Maria Losito[1,6]
Elio Conte[1,†]
and Franco Orsucci[7]
[1]School of Advanced International Studies on Applied Theoretical and Non Linear Methodologies of Physics, Bari, Italy
[2]Clinic Villa Miralago, ASL Varese, Italy
[3]Department of Pharmacology and Clinical Pharmacology, Faculty of Medicine, Istinye University, Istanbul, Turkey

---

* This is a reformatted version of an article previously published in *Chaos and Complexity Letters*, Volume 10 Issue 3 (2016), New York: Nova Science Publishers.
† Corresponding Author's E-mail: elio.conte@fastwebnet.it.

In: Complexity Science in Human Change
Editors: Franco F. Orsucci and Nicoletta Sala
ISBN: 979-8-89113-408-9

[4]Hunan Agricultural University,
Computing and Information Science and Statistics China, Changsha
[5]Department of Basic Sciences and Human Oncology, University of Bari, Bari, Italy
[6]Clinic Santa Maria, G.V.M. Hospital, Bari, Italy
[7]Psychoanalysis Unit, University College, London, England

Abstract

We demonstrate that Rorschach ink-blots are multifractals. Normal subjects have been studied by recording their GSR, a neurological and psychophysiological signal, during two minute's observation of each Rorschach ink blot. We have determined that the induced perceptive dynamics in the subjects are also multifractal. Once studied within the framework of the theory of complexity, we have found that the intrinsic complexity of the ink-blots induces in the subjects an emotive and cognitive condition of profound inner conflict that may be estimated by the generalized Hurst exponent of the subject, as arising from the GSR recording. We expect that the results are important not only in the theory of complexity but also in the clinical study of cases of neurological, psychological and psychiatric interest.

Keywords: theory of complexity, multifractals, computational neuroscience, Rorschach ink-blots, Galvanic Skin Response (GSR)

1. Introduction

In recent years the science of the complexity and, in particular, the connected application of fractal analysis has been extensively used to investigate phenomena relevant to neuroscience (Smith Jr et al., 1996; Fernández and Jelinek, 2001; Jelinek et al., 2004a, 2004b; Cornforth and Jelinek, 2005; Losa, 2014, 2015; Mattei, 2014; Di Leva et al., 2015). This kind of analysis has proven especially valuable for investigating various cell types of interest in neuroscience (Jelinek and Steinke, 1996; Nonnenmacher et al., 1994; Ristanović et al., 2002, Ristanović and Losa, 2013; Smith Jr et al., 1996). In essence, fractal analysis examines the scaling inherent in a pattern or dataset and assigns a number to the scaling (Karperien A, 2013; Gisiger, 2001). The estimated number is a fractal dimension (FD), a statistical index of complexity (Mandelbrot, 1983; Schroeder, 1991; Iannaccone and Khokha, 1995; Fu-Zen Shaw, 1999; Nazahah et al., 2012). Authors have focused their activity on FDs obtained from patterns extracted from digital images using a fractal analysis approach known as box-counting, in which the derived FD is called the box-counting dimension (BD) (Block et al., 1990; Jian et al., 2009). A single, global FD or BD index describes a pattern that is consistently self-similar or scale-invariant everywhere (i.e., monofractal), but neither uniquely nor completely describes the pattern. Indeed, sometimes the scaling within a dataset may itself vary (Wang et al., 2014a, 2014b, 2015). In essence, we have a non-uniform scaling determining a multifractal that may be quantified accordingly by measuring multiple dimensions (Mandelbrot, 1983; Voss and Wyatt, 1991; Stanley et al., 1999). The theoretical basis of multifractality has been reviewed in depth by several authors and will not considered here in detail (Chhabra and Jensen, 1989; Block et al., 1990; Vicsek, 1992; Jestczemski and Sernetz, 1996). Multifractal analysis has been applied in neuroscience also, even if analysis of multifractality of biological images remains a rather difficult matter often marked by some limitations (Nonnenmacher et al. 1994; Jelinek and Fernández, 1998; Fernández et al., 1999; Jelinek et al., 2004a; Bullmore et al., 2009). In order to allow chaos and fractal analysis of images, one of us, in collaboration with other authors (Conte et al., 2006; Conte et al., 2008a; Conte et al., 2012; Conte et al., 2009; Conte et al., 2008b) started studies years ago in which we realized for the first time a new method of elaboration where images were selected and converted into the three basic colors Red/Green/Blue (RGB) (or, perhaps, gray scale) for each pixel, thereby obtaining a pixel-space series, suitable for fractal and chaos analysis. Following the advent of the Detrended Fluctuation Analysis (DFA) (Peng et al., 1994) for the time series, multifractal detrended fluctuation analysis (1D-MF- DFA) was introduced in 2002 (Kantelhardt et al., 2002) and the method was extended also to the two-dimensional case (2D MF- DFA) (Gu and Zhou, 2006). A 2D MF-DFA was articulated by one of us and his coauthors (Wang et al., 2014a; Wang et al., 2014b; Wang et al., 2015).

The inner structural characteristics of biomedical electrophysiological signals are often interpreted and analyzed by traditional methods, but it has become progressively clearer in the studies of the last ten years that the basic inner structure cannot be captured by such conventional measures, since biomedical signals stemming from a wide range of physiological phenomena possess a scale invariant structure. Generally speaking, a biomedical signal has a scale invariant structure when the structure repeats itself on subintervals of the signal. Formally, the biomedical signal X(t) is scale invariant when $X(ct)=c^{H} X(t)$. Consequently it is of basic interest to examine such scale invariant behavior by using a Fractal Analysis that estimates the power law exponent, H, which defines the particular kind of scale invariant structure of a specific biomedical signal. Fractal analyses are presently employed in biomedical signal processing to define the scale invariant structure in ECG, EEG, MR, and X-ray pictures (Lopes and Betrouni, 2009). Considerable interest has been demonstrated in the analysis of the scale invariant structures of the inter-spike-interval of neuron firing, inter-stride-interval of human walking, inter-breath-interval of human respiration, and inter-beat intervals of the human heart deriving important factors differentiating between healthy and pathological conditions (e.g., Ivanov et al., 1999; Peng et al., 2002; Zheng et al., 2005; Hausdorff, 2007), and between different types of pathological conditions (e.g., Wang et al., 2007). Scale invariant structures have also been found in spatial phenomena like the branching of the nervous system, lungs (e.g., Bassingthwaighte et al., 1990; Abbound et al., 1991; Weibel, 1991; Krenz et al., 1992), and bone structure (Parkinson and Fazzalari, 1994). The results differentiate between healthy and cancerous tissues (Atupelage et al., 2012). Recently we have also conducted studies of neuron locations, the neural network, calcium activity and connectomics in the C-Elegans (Conte 2016a and Conte 2016b). This work provided further neurological confirmation that brain functions demonstrate scale invariant structure, a structure evident within even the most elementary neural network as represented in a small creature such as the C-elegans that is structured by a very restricted number of neurons. Several results in the last decade have suggested that changes in the scale invariant structure of biomedical signals reflect changes in the adaptability of physiological processes, implying successful treatment of pathological conditions might change fractal structure and improve health (Goldberger, 1996; Goldberger et al., 2002). Fractal analyses are therefore promising prognostic and diagnostic tools in biomedical signal processing.

It is not surprising that in this framework, fractal and multifractal analysis of images relating the individual neuron, groups of neurons and their interconnections assume a particularly valuable interest in neuro-science.

While at the forefront of new and advanced methods of investigation, it must be outlined on the other hand, that unfortunately, this kind of analysis is not exempt from difficulties, particularly when the aim of the investigation is to more deeply understand the inner mechanisms and the temporal dynamics of these types of cells.

The present paper represents subsequent work following the publication (Conte et al., 2008a) in which we introduced a new methodology, the variogram, aimed at calculating the Generalized Fractal Dimension, the Fractal Measure, the Fractal Variance Function, and the Marginal Density functions of a given spatial (time) series in the sphere of biological dynamics, with particular interest in the analysis of images of biomedical or psychological interest. For details regarding this methodology and for detailed previous studies on the general formulation of the method, here including also Rorschach ink-blot fractal structure, see the references (Conte et al. 2008a, 2008b, 2009a, 2012a).

It is well known that the Rorschach Inkblot Test is a projective psychological test of personality, in which a subject's interpretations of ten standard, abstract and ambiguous designs (ink-blots), given in sequential order, are then analyzed, yielding information about the emotional state of the subject, his intellectual functioning, and integration. Rather recently (Conte 2016c, 2015a, 2015b, 2015c, 2011a, 2011b, 2011c, 2013a, 2013b, 2011d, 2009b, 2012b, 2007, 2010, 2009c, 2008c), we performed theoretical studies and experimentation also on Gestalt and ambiguous figures, and the results enabled us to introduce a quantum like model of the perceptual mechanism with relation also to the dynamics of how the human visual system resolves perceptual ambiguity in stimuli that contextually may induce multiple potential interpretations.

The aim of our studies is to investigate the possible implications of existing time or space scale invariant structure at the level of perception, cognition and the structure of our mental states. It is well known that E.H. Weber and G.T. Fechner, researchers that were active in the years 1885-1887, were the first to introduce physical methods in psychology since they analyzed in detail the mechanism of perception, and attempted to find a relationship between the physical magnitudes of the given stimuli and the perceived intensity of these stimuli during the perception of the subject. In brief, they were the first to consider the possibility of linking a stimulus of physical nature with the corresponding human-subjective response to it. Their results may be expressed by the following equation:

$$p = kLnS + C$$

where p relates perception whose stimulus is given by S. C and k are constants to be determined by the experimental framework.

We recently located another important result: that being the work of O.V. Mitina and F.D. Abraham [http://www.blueberry-brain.org/dynamics/mitina-fractal-perception.htm] who studied the use of fractals in perception. In this case fractals were chosen as stimuli, and they were submitted to an observer's perception at increasing levels of complexity. Some stimuli evoked subjective sensations including the subjective perception of time during fractal observation, the subjective assessment of complexity and the aesthetic attractiveness of the visual objects which were experimentally measured. It was demonstrated that the evoked subjective sensations were strongly determined by the fractal dimension of the fractal objects.

In the field of perception research, it has been found that many of our sensory systems follow psychophysical (psycho-physiological) power laws that are self-similar, because the same relationship holds between the variables independent of how they are scaled or rescaled. Note that in every case we always have a relation mediating internal subjective quality that defines our conscious experience, through an external physical quantity of particular physical intensity. This result seems to open novel perspectives of innovative investigation not only in the general fields of perception and cognitive science, but also in the most general tentative attempt toward a further understanding of the dynamics of our mental entities. As an example, an interesting result established the fact that (Conte et al.1991) a power law relates word rank and word frequency in many natural languages. If we list the words in a book by rank order of word popularity with respect to the number of times each word appears, we find a power law relationship. Analysis of language under different experimental conditions, also including the interactive relationship between the subject and the therapist was recently studied by us (Orsucci et al. 2013), demonstrating encouraging results under this new direct line of investigation. Under the psychological and psychiatric profile, it seems that important features relate to the fact that fractals delineate systemic dynamics, since lower dimensional objects strive towards higher dimensions by recursively adding structure, or instead, they start as higher dimensional objects that retreat progressively into lower dimensions by recursive removal of the structure. Also according to Terry Marks-Tarlow (http://goertzel.org/dynapsyc/2002/ObserverObserved.htm), in the psycholo-gical dynamics of our mental entities, we may observe a distinction between progressing towards higher versus regressing towards lower dimensionality as a model, depending upon whether a subject's psyche is better characterized by structure building evolution of consciousness, or structure eroding involution. Along with such important new intuitions we have to link the fact that, as previously stated, fractal structures have been found to be ubiquitous in the brain, and, in particular, fractal behaviors have been found also related to specific mental entities e.g., in our mood shifts. One of the most important results is that fractal dynamics have been identified (Delignières et al., 2004) also in tests of self-esteem and self. Following such a line of analysis, one is tempted to retain that consequently fractals and multifractals should also exist within the inner structure of our personality, since it organizes itself in self-similar patterns of behavior at different scales of observations. It has also been suggested that the tiniest fragment of a dream has the capacity to reflect the whole of the psyche so that every dream carries out a fractal structure. From the micro-level of speech patterns, through a medium scale event of a chance encounter on the highway, to the large scale level of ongoing business relations, people generally demonstrate self-similar behavior across multiple scales of observation. When this gets rigidly stereotyped, one may think in terms of Freud's notion of repetition compulsion.

In conclusion we retain that a new perspective within studies is emerging that may be of interest in neurology, in psychology and psycho-physiology, as well as in psychiatry. The general reason is that it seems that the concept of self and of identity could be re-conceptualized in multivalent terms, that is to say, in terms of self-similar processes repeating on multiple size and event scales. As has been demonstrated, different studies seem to show that fractal structures affect human psychology and, in particular, enforce the position that self-perception is an emergent product of a dynamical system based on fractal processes and composed of multiple interacting components. Another reason is also that, according to the studies of M. Schroeder in 1991 (Schroeder 1991), fractals exist symbolically, and this definitively opens a new pathway of investigation into the nature of the mind since, consequently, fractals should exist in the psyche at symbolic, invisible realms (Norman, 2016).

These results give support to our present paper. As said, fractals exist symbolically and thus fractals could exist in the psyche in symbolic, invisible (unconscious) realms, and since, as discussed in particular in (http://goertzel.org/dynapsyc/2002/ObserverObserved.htm), the fractal dimension should be a determinant in the characterization of the subjective evoked sensation or emotions elicited during image perception. Rorschach's ink blots were extensively used in the early years of psychological and psychiatric studies, but in later years their use has progressively decreased and is often dismissed by modern scholars. The initial assumption was that they were able to induce overt responses in a subject in correspondence with his projected story, indicating his emotional state, his intellectual functioning and so, revealing vital data on this basis, its use had initially become of basic importance in psychological and psychiatric analysis. However, a basic limit was revealed in the absence of a firm, standardized interpretive basis of test results.

Of course some points remain clear. Rorschach ink blots do not have a defined meaning and require subjective identification when perceived and interpreted by the subject. At the same time, as previously studied and analyzed by us (Conte 2008a) they demonstrate substantial attractivity mixed with a large measure of complexity. In research the term complexity should not be considered as a generic or philosophical term. Today we have the science of complexity emerging from the intrinsic inner nonlinear structure of natural processes, and it may be quantified just as well as any other physical or physiological variable. Of course, often complexity is strongly related to the fractal or multifractal structure of the investigated process. The presence of high complexity in the Rorschach ink-blots determine in the perceptive and cognitive phases of the subject observing the images, a situation involving many coexisting interpretive alternatives in a kind of quantum model, such as is demonstrated in (Conte 2011a, 2011b, 2011c, 2013a, 2013b; Conte et al.2016c, 2015a, 2015b, 2011d, 2009b, 2012b, 2007, 2010, 2009c, 2008c). This situation involving very complex coexisting multiple alternatives of interpretation, induces in the subject a substantial condition of conflict. To explain this situation of conflict through our internal quantum model let us use an example. Consider a task posed to a subject: "do you like this picture?" During the actions of perception and cognition when the subject is looking at the figure, a simultaneous coexistence of alternatives define the mental state of the subject. It is resolved and represented by a mental function that mathematically may be represented in a superposition of probabilistic alternatives

$$|Mental\ State\ of\ the\ subject\rangle = c_1|I\ like\rangle + c_2|I\ do\ not\ like\rangle$$

$|c_1|^2$ and $|c_2|^2$

define the probabilities that finally, the subject will select only one of the alternatives. It is easily understandable that when the two probabilities have different values corresponding to the inner mental dynamics of the subject (as example 0.9 and 0.1), the inner conflict of the subject in relation to the final decision will be very moderate, while instead, if one probability is 0.51 and the other 0.49, the inner conflict will be high. In the case of the Rorschach ink blots the subject is not called to select between two alternatives as in the previous simple example. Owing to the high complexity of the posed ink-blots he/she will determine in themself a large number of similar alternatives and thus, he/she will be in a great condition of inner conflict. Such inner conflict will determine in the subject a situation based in memory that will be linked to his/her projected story. He/she will be able to decide and give an answer on the basis of their conflicting ambivalent situation, emotional state, intellectual functioning and his/her affective/cognitive level of integration. Emotional responses and connected cognitive engagement are always associated with the previous experiences of a subject, particularly relating to childhood and his/her relation to parental figures. The complexity of the relation will determine the inner situation of conflict among simultaneous alternatives and complexity, which should, we believe, be related to the multifractal structure of the images used. Therefore, we have derived two questions that deserve consideration. 1) Do Rorschach's ink blots really exhibit a multifractal structure? 2) When Rorschach's ink blots are given as the perceptive input to a subject, may we use his/her measured psycho-physiological parameters in clinical applications?

We know that GSR has definite neurological correlates (Norman et al., 2016a-b). Consequently, if points (1) and (2) could be firmly established, we will have derived a preliminary approach of basic importance in neurology and psychiatry.

2. Materials and Methods

The details of the method have been reported elsewhere (Gu and Zhou 2006; Wang C. et al., 2014; Wang et al., 2015) and the reader is invited to examine all the basic features of the method by reading the original works.

In essence, the method consists of five steps of analysis that are reported in Appendix A.

In accordance with the instructions of the International Association for Rorschach and Projective Tests (http://www.rorschach.com/index.php?id=2) the ten colored Rorschach cards were downloaded and saved in a computer as .jpg files. The images were explored through their treatment and analysis consisting in a partition of the visible surface into a $M_S x N_S$ non-overlapping square sub-surface of equal lengths and treated on the basis of the procedure indicated in Appendix A, and reassumed in five steps.

Ten normal subjects were selected within the age range of 20-45 years. All the subjects expressed their formal consent to participate in the experiment after detailed explanation of its content. All the experimental investigation was performed at our laboratory in Casarano (Lecce, Italy) under the direction of one of the authors (E.C.) and the direct collaboration and assistance of the specialized authors of the present paper.

Psychotherapeutic and psychiatric scholars previously interviewed the subjects and, subjects were selected who had no psychological disorders or other psychiatric pathologies, and excluding also, subjects having high blood pressure, cardio-circulatory insufficiency, diabetes or other chronic diseases. In particular, subjects who demonstrated heart related disease or coronary artery disease (CAD), heart failure, or irregular heartbeat, were excluded. We dismissed subjects having abnormal Q wave, WPW syndrome. Subjects with hypertrophy on the left ventricles of the heart or heart hypertrophy were dismissed. Subjects having less than 90% of normal RR interval were also dismissed. Subjects who smoked or used alcohol were excluded. No food within one hour prior to the exam was permitted. The test was performed at 10.30 a.m. Recording was performed with the subject sitting comfortably, under soft and distant indoor light and appropriate indoor temperatures (20°C - 23°C) and, in a condition of absolute calm without noise or disturbances. Galvanic Skin Response (GSR) recording in real time including of course tone and phase values, was performed by using an Elemaya device with two Ag-electrodes connected respectively to the index and ring fingers of the left hand. No conversation was allowed during the recording. The exam was performed with the subject's eyes open. Rorschach cards were shown to the subjects using a 15-inch computer monitor without any image distortion, shown in temporal sequence at preselected intervals of one minute's separation. The subject was given the time of two minutes to look at each picture. During the observation of the picture no issue or question was posed to the subject so as to evaluate his perceptual activity and cognitive processing through the real time GSR recording only, and *not* through verbal expression or spoken conceptual elucidation or enquiry. Tone and phase of each GSR recording were stored in the computer interfaced with the device. For each GSR recording the latency, the rise time, the Peak value and the Half Time Recovery as well as the mean values of the GSR Tone, were estimated by our methods of analysis. We will publish the results in a separate publication.

The time series of the values of the GSR Tone and Phase obtained for each figure and for each subject were subsequently submitted to 1-D-DFA multifractal analysis. This is the one dimensional version of the analysis reported in Appendix A.

To condense: we first performed the multifractal analysis of the ten Rorschach ink blots and subsequently, following a procedure derived from years of work in our laboratory, we examined the psycho-physiological monitoring, treatment and analysis of the electrophysiological GSR signal derived from each subject when presented with each figure while allowing his/her personal, subjective responses and cognition to develop without providing any verbal information.

3. Results

First let us examine the results that we obtained in analysis of the Rorschach ink blots.

Let us verify that we have really found that the images represent a multifractal structure. The results relating $h(q)$, $\tau(q)$, α and $f(\alpha)$ are given in Figures 1- 4 respectively.

As expected $h(q)$ shows monotonically decreasing values as function of q. $\tau(q)$ demonstrates its typical nonlinear dependence upon q, and $f(\alpha)$ has the typical distribution of a multifractal spectrum.

In conclusion, we have shown that the Rorschach ink-blots used in psychology and psychiatry as projective structures not having a definite meaning, really do possess a complex multifractal structure.

Table 1. Estimated Values of $h(q)$

h(q)										
q	RORS 1	RORS 2	RORS 3	RORS 4	RORS5	RORS6	RORS7	RORS8	RORS9	RORS10
-10	2,122	2,539	2,986	2,738	2,490	2,571	2,190	2,067	2,069	2,226
-9	2,107	2,517	2,955	2,717	2,478	2,550	2,169	2,059	2,060	2,204
-8	2,093	2,491	2,917	2,691	2,465	2,525	2,146	2,052	2,052	2,178
-7	2,079	2,459	2,866	2,659	2,450	2,494	2,121	2,044	2,044	2,150
-6	2,065	2,420	2,798	2,619	2,430	2,455	2,095	2,036	2,036	2,119
-5	2,052	2,371	2,704	2,567	2,402	2,404	2,070	2,029	2,028	2,088
-4	2,039	2,309	2,572	2,498	2,361	2,336	2,049	2,022	2,021	2,060
-3	2,028	2,233	2,392	2,402	2,294	2,245	2,032	2,016	2,015	2,039
-2	2,018	2,149	2,193	2,273	2,191	2,141	2,020	2,011	2,009	2,023
-1	2,009	2,078	2,070	2,127	2,090	2,063	2,012	2,007	2,005	2,012
0	2,002	2,034	2,024	2,034	2,034	2,025	2,007	2,003	2,001	2,004
1	1,995	2,012	2,008	2,003	2,013	2,010	2,003	2,000	1,998	1,999
2	1,990	2,001	2,001	1,996	2,005	2,004	2,001	1,998	1,996	1,995
3	1,986	1,996	1,998	1,995	2,003	2,002	1,999	1,997	1,994	1,992
4	1,983	1,994	1,997	1,995	2,002	2,001	1,999	1,996	1,993	1,991
5	1,980	1,993	1,996	1,996	2,002	2,001	1,998	1,995	1,992	1,989
6	1,979	1,993	1,996	1,997	2,002	2,001	1,998	1,994	1,991	1,988
7	1,977	1,993	1,996	1,997	2,002	2,001	1,998	1,994	1,991	1,988
8	1,977	1,993	1,997	1,998	2,003	2,001	1,998	1,994	1,990	1,987
9	1,976	1,993	1,997	1,999	2,003	2,001	1,998	1,994	1,990	1,987
10	1,976	1,994	1,997	1,999	2,003	2,001	1,998	1,994	1,990	1,987

Table 2. Estimated Values of $\tau(q)$

τ(q)										
q	RORS 1	RORS 2	RORS 3	RORS 4	RORS5	RORS6	RORS7	RORS8	RORS9	RORS10
-10	-23,216	-27,394	-31,855	-29,375	-26,895	-27,706	-23,898	-22,666	-22,686	-24,255
-9	-20,966	-24,657	-28,599	-26,449	-24,305	-24,947	-21,524	-20,534	-20,542	-21,833
-8	-18,743	-21,929	-25,334	-23,528	-21,722	-22,196	-19,170	-18,414	-18,415	-19,427
-7	-16,551	-19,214	-22,062	-20,616	-19,147	-19,454	-16,846	-16,309	-16,305	-17,048
-6	-14,389	-16,519	-18,787	-17,717	-16,577	-16,727	-14,568	-14,218	-14,213	-14,713
-5	-12,258	-13,855	-15,519	-14,837	-14,012	-14,019	-12,350	-12,145	-12,140	-12,440
-4	-10,157	-11,238	-12,289	-11,991	-11,446	-11,343	-10,195	-10,089	-10,084	-10,241
-3	-8,084	-8,699	-9,176	-9,205	-8,881	-8,734	-8,096	-8,048	-8,044	-8,115
-2	-6,036	-6,298	-6,387	-6,545	-6,383	-6,282	-6,040	-6,022	-6,019	-6,045
-1	-4,009	-4,078	-4,070	-4,127	-4,090	-4,063	-4,012	-4,007	-4,005	-4,012
0	-2	-2	-2	-2	-2	-2	-2	-2	-2	-2
1	-0,005	0,012	0,008	0,003	0,013	0,010	0,003	0,000	-0,002	-0,001
2	1,980	2,002	2,002	1,991	2,011	2,008	2,002	1,997	1,992	1,990
3	3,958	3,989	3,994	3,984	4,008	4,005	3,998	3,990	3,983	3,977
4	5,931	5,976	5,987	5,980	6,008	6,004	5,994	5,983	5,971	5,962
5	7,901	7,966	7,981	7,979	8,010	8,003	7,990	7,974	7,959	7,946
6	9,871	9,958	9,977	9,980	10,012	10,004	9,986	9,966	9,947	9,930
7	11,841	11,951	11,974	11,981	12,016	12,005	11,984	11,958	11,935	11,914
8	13,813	13,945	13,972	13,984	14,020	14,007	13,981	13,951	13,922	13,898
9	15,787	15,941	15,971	15,987	16,024	16,009	15,980	15,944	15,911	15,882
10	17,763	17,937	17,970	17,990	18,029	18,011	17,979	17,938	17,900	17,868

Table 3. Estimated Values of $\alpha(q)$

α(q)										
q	RORS 1	RORS 2	RORS 3	RORS 4	RORS5	RORS6	RORS7	RORS8	RORS9	RORS10
-7	2,177	2,696	3,265	2,900	2,574	2,729	2,288	2,097	2,101	2,341
-6	2,147	2,665	3,247	2,878	2,570	2,706	2,240	2,081	2,083	2,291
-5	2,117	2,615	3,183	2,839	2,560	2,663	2,188	2,065	2,066	2,232
-4	2,089	2,536	3,040	2,766	2,525	2,584	2,136	2,050	2,049	2,170
-3	2,064	2,431	2,827	2,643	2,451	2,470	2,091	2,036	2,035	2,114
-2	2,041	2,311	2,581	2,485	2,348	2,337	2,055	2,023	2,022	2,068
-1	2,022	2,194	2,344	2,318	2,234	2,208	2,030	2,013	2,012	2,035
0	2,006	2,098	2,156	2,170	2,127	2,103	2,014	2,006	2,004	2,013
1	1,993	2,034	2,045	2,064	2,049	2,036	2,005	2,000	1,998	2,000
2	1,984	2,002	2,004	2,009	2,011	2,007	2,000	1,996	1,994	1,992
3	1,978	1,991	1,995	1,996	2,001	2,000	1,997	1,994	1,991	1,988
4	1,974	1,990	1,994	1,997	2,001	1,999	1,997	1,993	1,989	1,986
5	1,972	1,991	1,995	1,999	2,002	2,000	1,997	1,992	1,988	1,984
6	1,971	1,992	1,996	2,001	2,003	2,001	1,997	1,992	1,988	1,984
7	1,972	1,994	1,997	2,002	2,004	2,001	1,997	1,993	1,988	1,984

In Table 4 and Figures 5-6 we report the results obtained in estimation of the Hurst exponent, the generalized fractal dimension and the strength of multifractality.

The corresponding results are given in Tables 1-3.

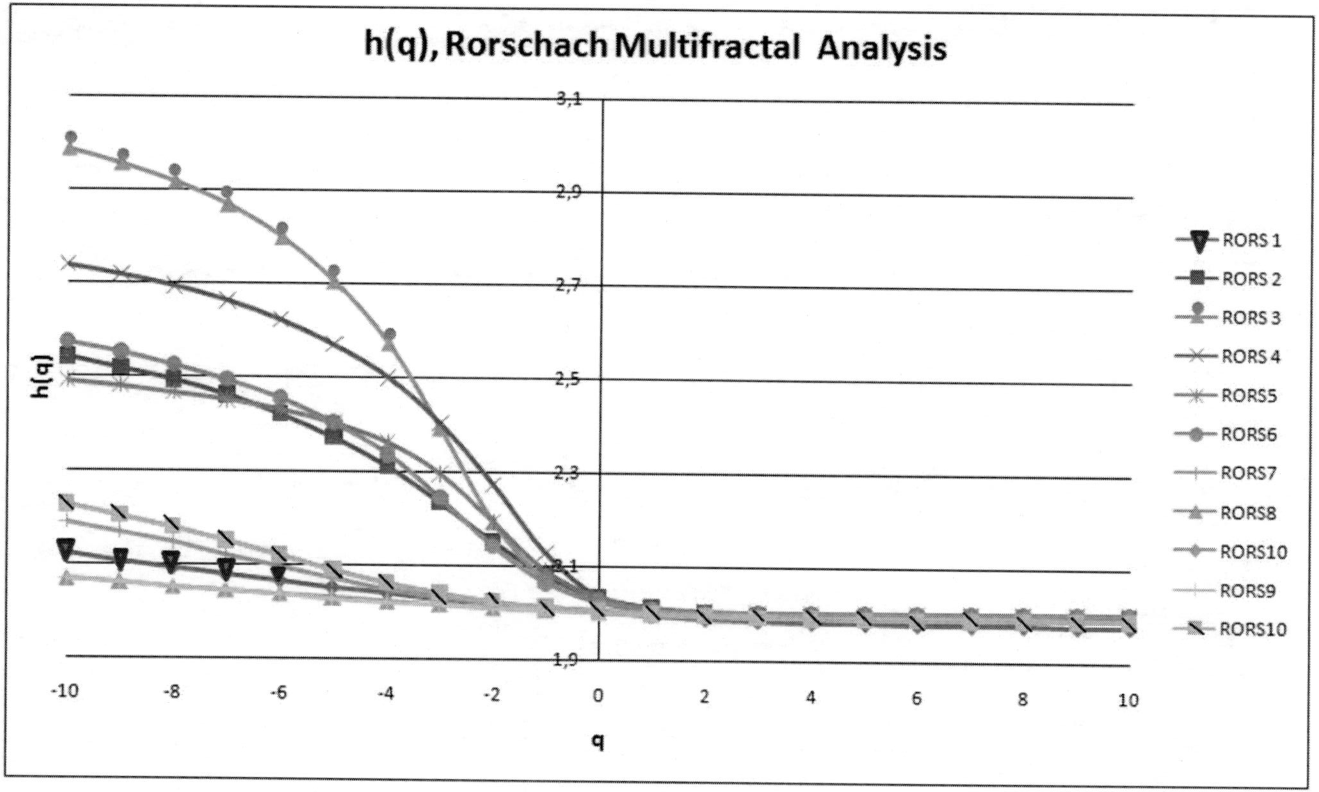

Figure 1. *h(q)* estimated values of the ten examined Rorschach ink-blots.

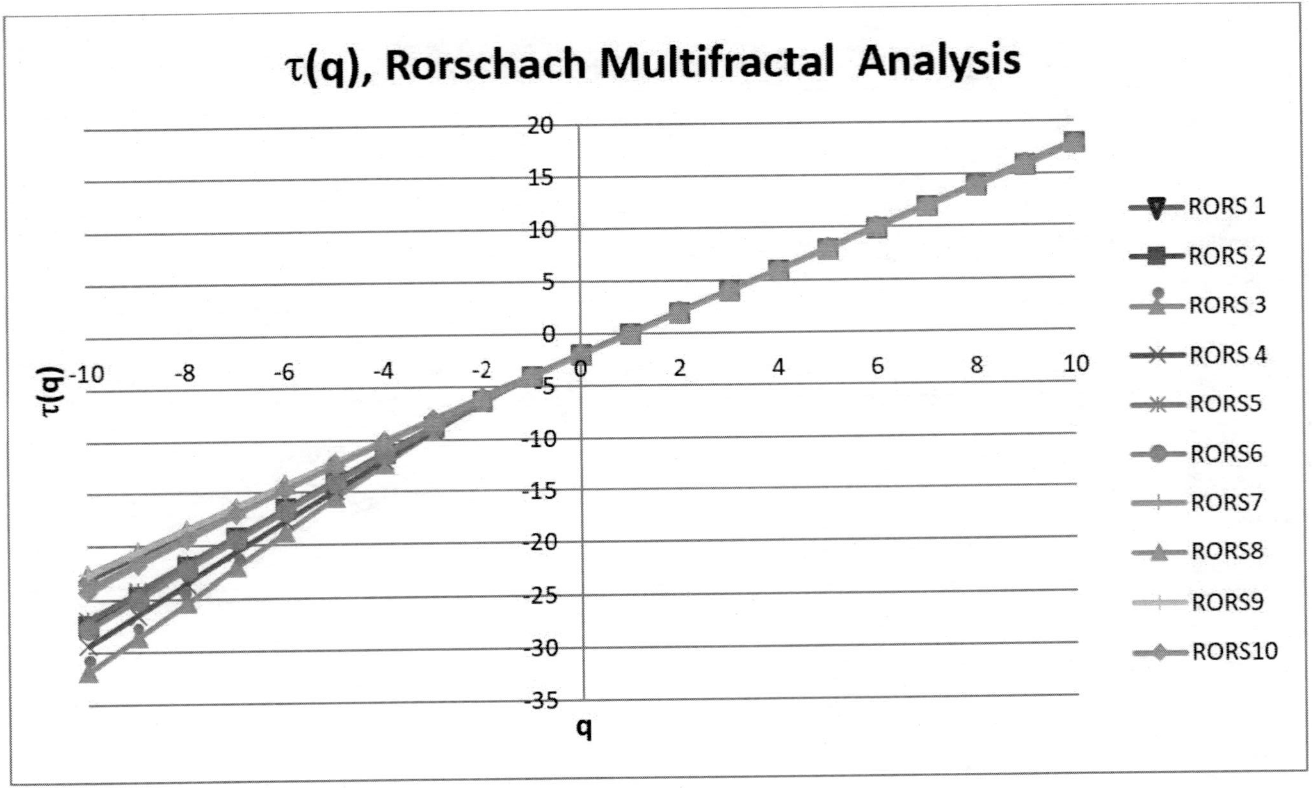

Figure 2. $\tau(q)$ estimated values of the ten examined Rorschach ink-blots.

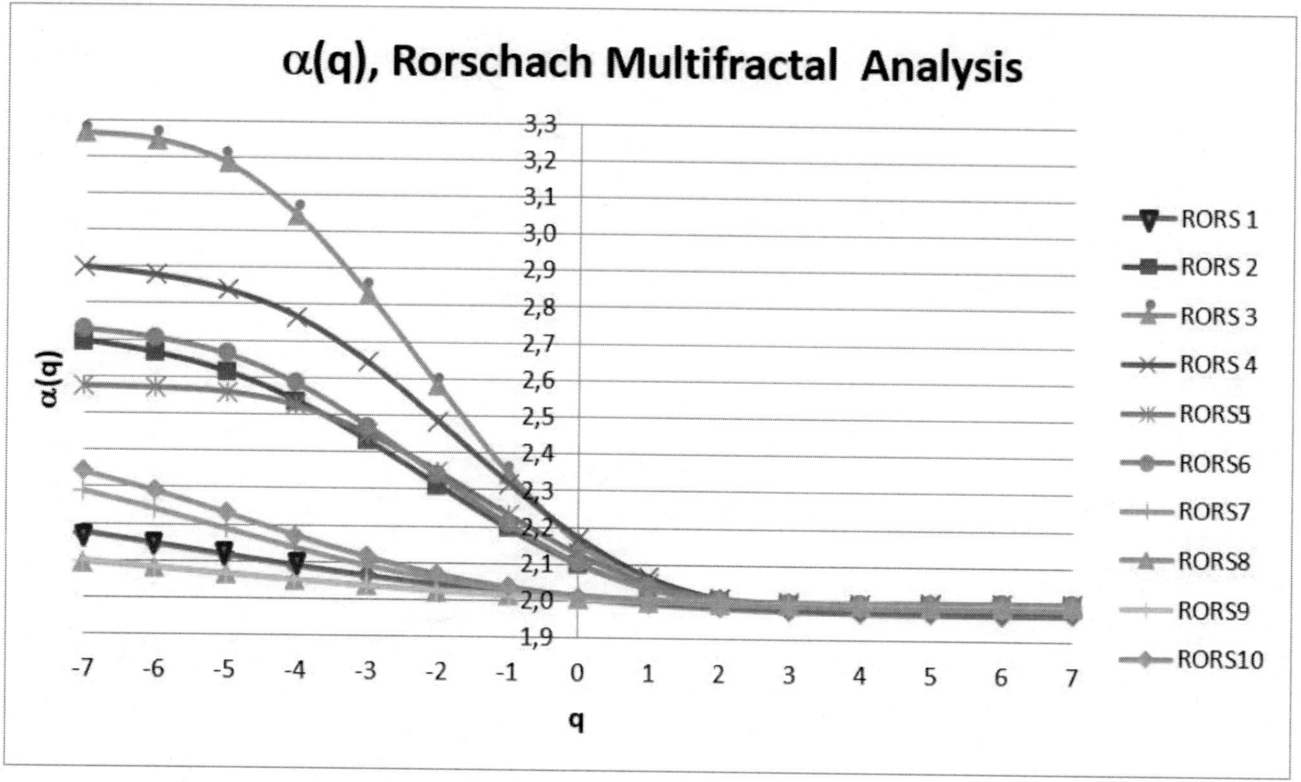

Figure 3. $\alpha(q)$ estimated values of the ten examined Rorschach ink-blots.

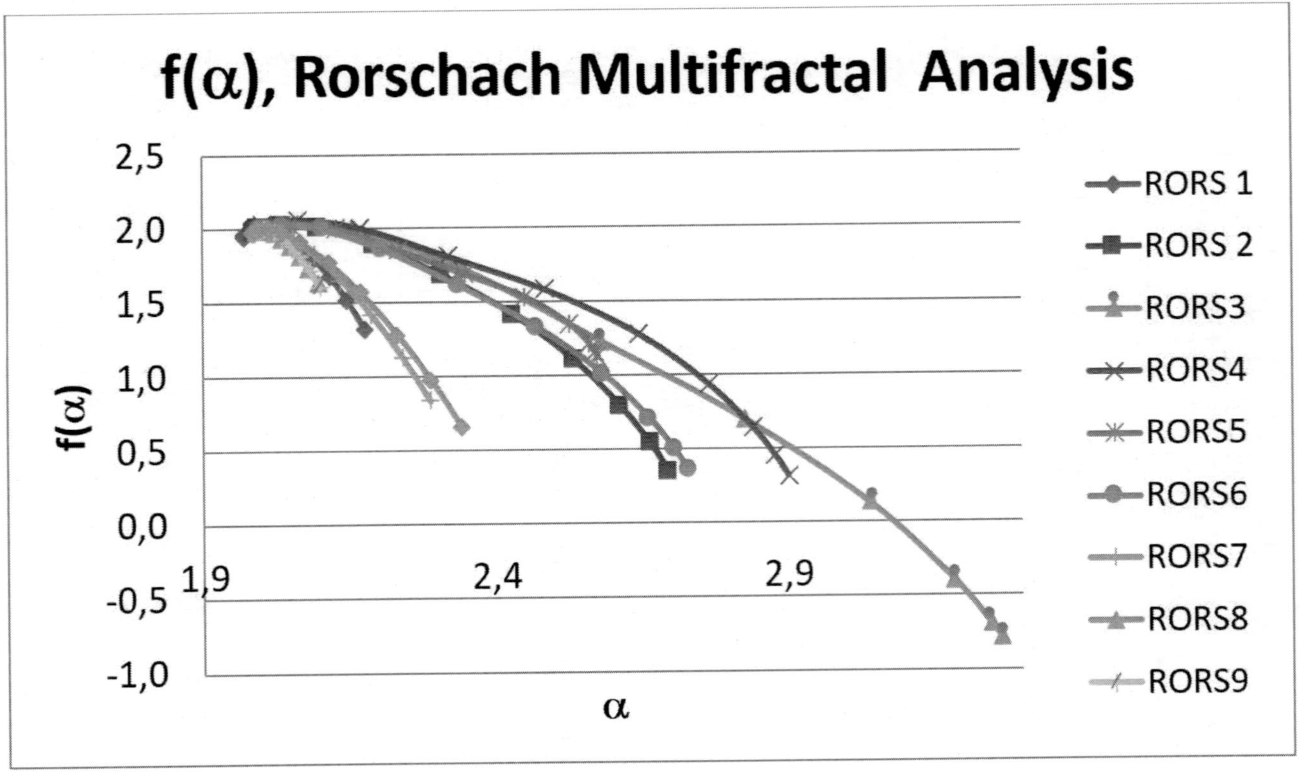

Figure 4. $f(\alpha)$ estimated values of the ten examined Rorschach ink-blots.

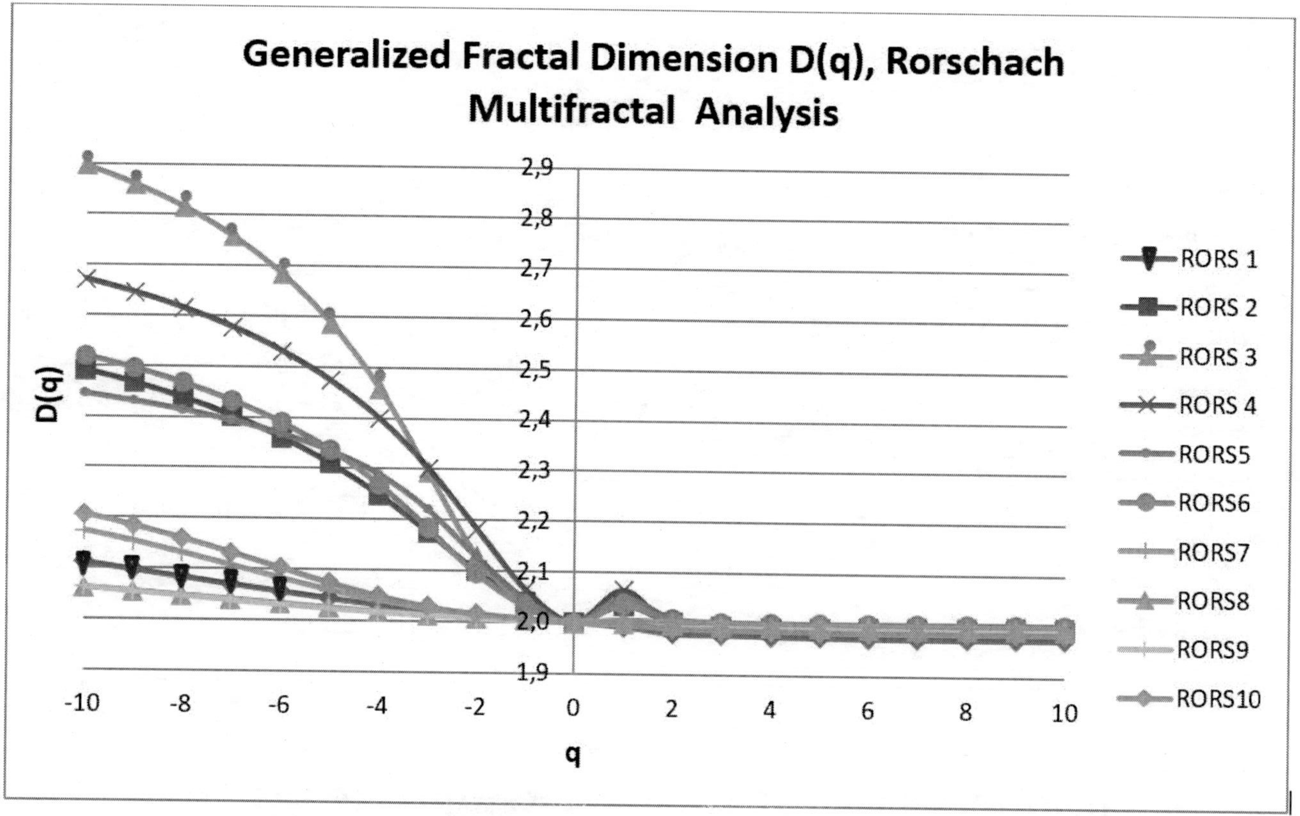

Figure 5. D(q) estimated values of the Generalized Fractal Dimension.

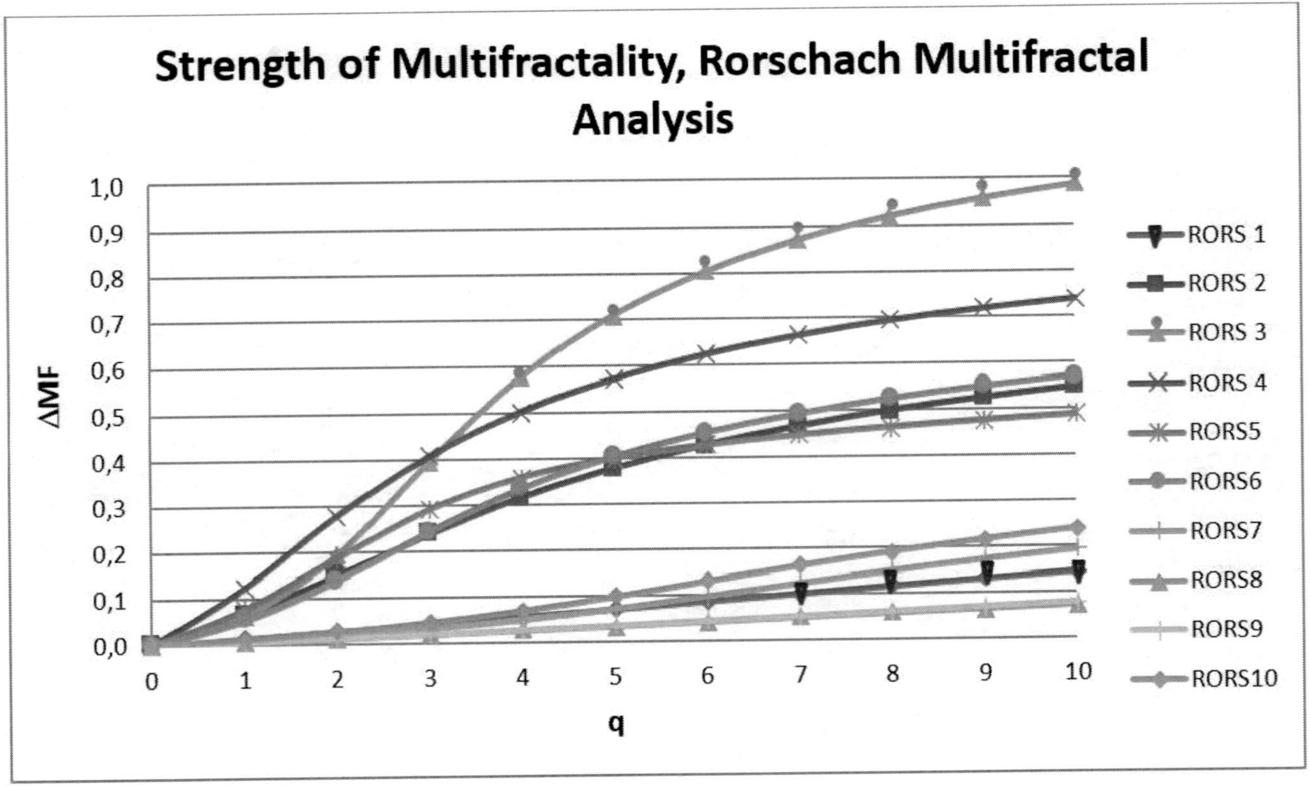

Figure 6. ΔMF estimated values, strength of multifractality.

The corresponding values are given in Tables 5 and 6.

Table 4. Hurst Exponents of the Rorschach Ink-Blots

Hurst Exponent									
RORS 1	RORS 2	RORS 3	RORS 4	RORS5	RORS6	RORS7	RORS8	RORS9	RORS10
1,990	2,001	2,001	1,996	2,005	2,004	2,001	1,998	1,996	1,995

Table 5. Estimated Values of the Generalized Fractal Dimension

D(q)										
q	RORS 1	RORS 2	RORS 3	RORS 4	RORS5	RORS6	RORS7	RORS8	RORS9	RORS10
-10	2,111	2,490	2,896	2,670	2,445	2,519	2,173	2,061	2,062	2,205
-9	2,097	2,466	2,860	2,645	2,431	2,495	2,152	2,053	2,054	2,183
-8	2,083	2,437	2,815	2,614	2,414	2,466	2,130	2,046	2,046	2,159
-7	2,069	2,402	2,758	2,577	2,393	2,432	2,106	2,039	2,038	2,131
-6	2,056	2,360	2,684	2,531	2,368	2,390	2,081	2,031	2,030	2,102
-5	2,043	2,309	2,587	2,473	2,335	2,337	2,058	2,024	2,023	2,073
-4	2,031	2,248	2,458	2,398	2,289	2,269	2,039	2,018	2,017	2,048
-3	2,021	2,175	2,294	2,301	2,220	2,184	2,024	2,012	2,011	2,029
-2	2,012	2,099	2,129	2,182	2,128	2,094	2,014	2,007	2,006	2,015
-1	2,005	2,039	2,035	2,063	2,045	2,032	2,006	2,003	2,002	2,006
0	2	2	2	2	2	2	2	2	2	2
1	1,993	2,034	2,045	2,064	2,049	2,036	2,005	2,000	1,998	2,000
2	1,980	2,002	2,002	1,991	2,011	2,008	2,002	1,997	1,992	1,990
3	1,979	1,994	1,997	1,992	2,004	2,003	1,999	1,995	1,991	1,989
4	1,977	1,992	1,996	1,993	2,003	2,001	1,998	1,994	1,990	1,987
5	1,975	1,992	1,995	1,995	2,002	2,001	1,998	1,994	1,990	1,987
6	1,974	1,992	1,995	1,996	2,003	2,001	1,997	1,993	1,989	1,986
7	1,974	1,992	1,996	1,997	2,003	2,001	1,997	1,993	1,989	1,986
8	1,973	1,992	1,996	1,998	2,003	2,001	1,997	1,993	1,989	1,985
9	1,973	1,993	1,996	1,998	2,003	2,001	1,997	1,993	1,989	1,985
10	1,974	1,993	1,997	1,999	2,003	2,001	1,998	1,993	1,989	1,985

Table 6. Estimated Values of the Strength of Multifractality

Strength of Multifractality										
q	RORS 1	RORS 2	RORS 3	RORS 4	RORS5	RORS6	RORS7	RORS8	RORS9	RORS10
0	0	0	0	0	0	0	0	0	0	0
1	0,014	0,066	0,063	0,123	0,077	0,053	0,009	0,006	0,007	0,013
2	0,028	0,148	0,192	0,277	0,186	0,137	0,019	0,013	0,013	0,028
3	0,042	0,237	0,394	0,407	0,291	0,243	0,033	0,019	0,021	0,046
4	0,057	0,315	0,576	0,503	0,359	0,335	0,050	0,027	0,028	0,070
5	0,071	0,378	0,708	0,572	0,401	0,403	0,072	0,034	0,036	0,099
6	0,086	0,427	0,802	0,623	0,428	0,454	0,097	0,042	0,044	0,131
7	0,101	0,466	0,870	0,662	0,447	0,493	0,123	0,050	0,053	0,162
8	0,116	0,498	0,920	0,693	0,463	0,524	0,149	0,058	0,062	0,191
9	0,131	0,524	0,959	0,718	0,476	0,549	0,172	0,065	0,070	0,217
10	0,145	0,546	0,989	0,739	0,487	0,570	0,192	0,073	0,079	0,239

All the results confirm multifractality. In addition, the other important result is that each analysis demonstrates that we have a net differentiation in the values for each figure. Consequently each image contains a detailed multifractal structure that is differentiated from the others and we consequently expect that this specialized multifractal structure of each ink blot, will induce different neurological as well as psychological reactions in the subjects during their direct observation. In Table 7 we give also the results of $\Delta\alpha$ and of Δf, as defined in the appendix A. Such basic indices indicate that we are in presence of images having elevated and differentiated roughness, irregularity and complexity associated with each ink-blot.

Table 7. Estimated Complexity in Each Rorschach Ink-Blot

	RORS 1	RORS 2	RORS 3	RORS 4	RORS5	RORS6	RORS7	RORS8	RORS9	RORS10
$\Delta\alpha$	0,206	0,706	1,270	0,904	0,574	0,730	0,291	0,104	0,113	0,357
Δf	0,648	1,661	2,796	1,712	0,882	1,653	1,168	0,357	0,380	1,314

According to the discussion elaborated in the introduction, we assert that such levels of roughness, irregularity and complexity have a decisive role in determining our condition of inner emotion and conflict, reaction and answer, when each figure is submitted to perception and cognition.

The first point of the present paper is thus established. Rorschach ink-blots are multifractals and their high and differentiated complexity is at the basis of the reactions that subjects manifest in normal as well as in pathological conditions, when these ink-blots are observed and also, to spontaneous or requested cognitive interpretation when they are used as a projective test.

In the context of this novel interpretation of the Rorschach ink-blots, it is important to show that the high values of multifractal and complex structure of the ink-blots induce a corresponding multifractal dynamic in the basic neurological and psycho-physiological GSR signal that is recorded when the subject observes each ink-blot, thus engaging his/her perception of each figure, while undisturbed by verbal interference or questions. Therefore, we may now disclose the results of the second section of our experiment, the psychophysiological GSR recording established during the direct perception of each figure.

We repeat that we applied the methodology of Appendix A but in the case of a 1-D-DFA time series applied to the data.

In order to explain the results let us consider the following premise. Let us consider three kinds of time series, the first being white noise, the second relating a monofractal time series and, finally, a multifractal time series. One of the possible salient features of the multifractal time series, with respect to the monofractal and white nose cases, is that in the multifractal case we may have small as well as large fluctuations in the values. As we know, the GSR recording relates values in the Tone that in the normal condition of the subject cover the range between 100 KOhm and 300 KOhm (10 microSiemens and 3.3 microSiemnes) with decreasing values responding to the particular emotive and cognitive condition of the subject. The phase in the GSR monitoring reveals important parameters as the baseline values, the latency time in respect to the reaction to a given inner or external stimulus, the rise time, the peak value as well as the half peak time value. These are parameters describing the sudden neurological and psychological changes in the mental and neurological states of the subjects. Every scholar of GSR monitoring knows, that during the GSR recording taken of a subject, large and rapid fluctuations in the values of the Tone and, in particular, of the Phases are observed in correspondence to rapid changes in the neurological and mental status of the subject, at each moment. Therefore, the 1-D-DFA analysis of the GSR is expected to give results only in correspondence to such large fluctuations.

This is precisely the result that we obtained in our experiments. First we obtained, and so confirmed, that the GSR time series recorded when the subject looking at each ink-blot is a multifractal for the tone and for the phase and, in particular, we discovered that such a multifractal regime is realized only for large fluctuations, that is to say, for sudden and relevant modifications in the neurological and mental state of the subject during his/her observation of each figure.

The results are reported in Figures 7 and 8 (and Tables 8, 9) for the GSR Tone and Phase respectively for $h(q)$. The values confirm that the subject responds to the visual inspection of ink-blots following neurological and mental multifractal dynamics.

All the normal subjects responded with very similar values with modest variations constrained within a range of 10%.

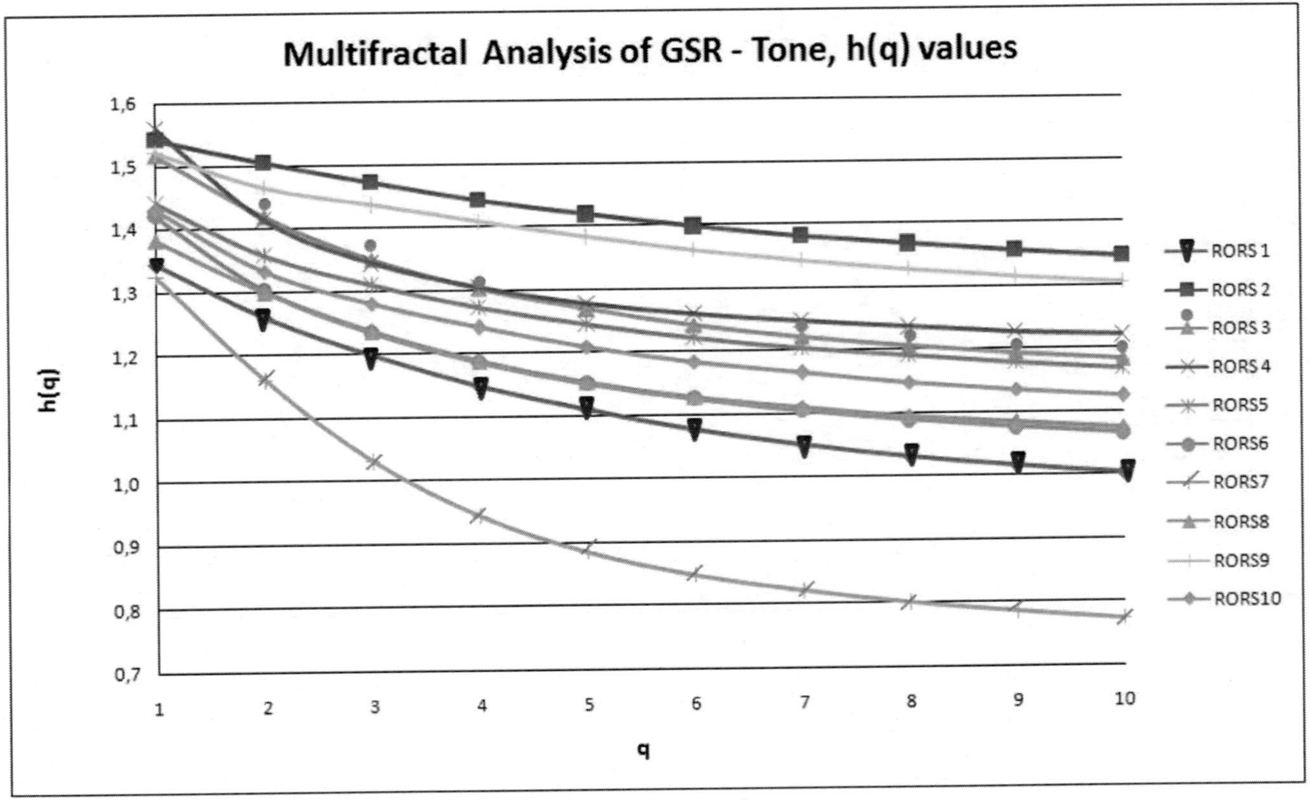

Figure 7. Multifractal GSR analysis, $h(q)$ - GSR-Tone values.

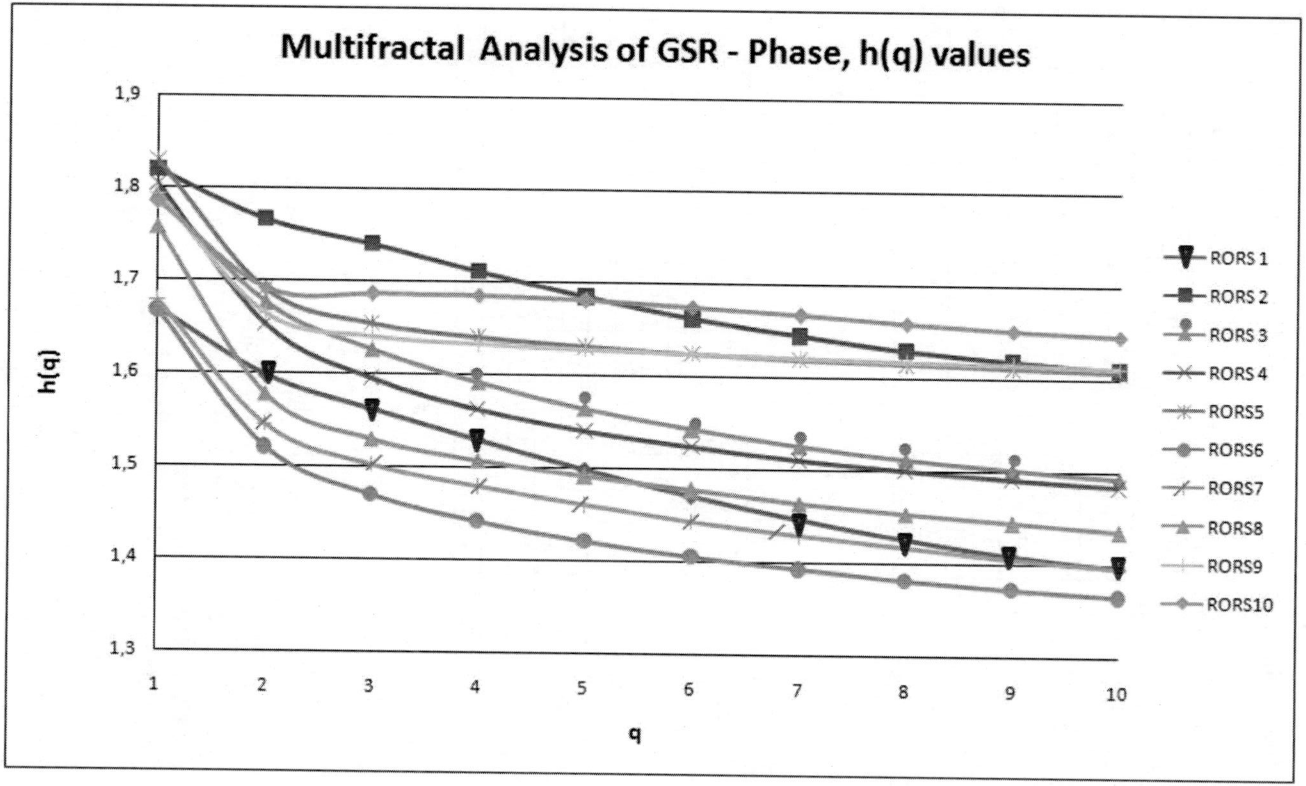

Figure 8. Multifractal GSR analysis, $h(q)$ - GSR- Phase values.

Table 8. Multifractal Analysis of the GSR -Tone, $h(q)$ Values

Multifractal Analysis of GSR - Tone, h(q) values										
q	RORS 1	RORS 2	RORS 3	RORS 4	RORS5	RORS6	RORS7	RORS8	RORS9	RORS10
1	1,343	1,542	1,516	1,559	1,439	1,419	1,324	1,382	1,520	1,430
2	1,258	1,503	1,417	1,412	1,356	1,300	1,158	1,297	1,463	1,331
3	1,197	1,470	1,350	1,343	1,308	1,232	1,030	1,234	1,434	1,278
4	1,148	1,441	1,302	1,302	1,272	1,185	0,942	1,186	1,406	1,239
5	1,108	1,416	1,267	1,276	1,243	1,150	0,885	1,152	1,381	1,208
6	1,076	1,396	1,240	1,258	1,221	1,123	0,846	1,127	1,359	1,183
7	1,051	1,379	1,220	1,245	1,203	1,103	0,820	1,109	1,341	1,163
8	1,031	1,366	1,204	1,234	1,189	1,088	0,800	1,094	1,326	1,147
9	1,016	1,354	1,192	1,226	1,177	1,075	0,785	1,083	1,314	1,134
10	1,003	1,345	1,181	1,219	1,168	1,065	0,773	1,073	1,303	1,123

Table 9. Multifractal Analysis of the GSR - Phase, $h(q)$ Values

Multifractal Analysis of GSR - Phase, h(q) values										
q	RORS 1	RORS 2	RORS 3	RORS 4	RORS5	RORS6	RORS7	RORS8	RORS9	RORS10
1	1,668	1,819	1,793	1,803	1,828	1,666	1,677	1,757	1,792	1,783
2	1,597	1,765	1,674	1,651	1,689	1,519	1,544	1,577	1,663	1,691
3	1,561	1,739	1,626	1,594	1,653	1,469	1,502	1,529	1,639	1,685
4	1,529	1,710	1,591	1,562	1,638	1,441	1,478	1,507	1,632	1,684
5	1,498	1,683	1,564	1,540	1,630	1,421	1,459	1,490	1,627	1,680
6	1,469	1,661	1,542	1,523	1,624	1,405	1,443	1,476	1,623	1,673
7	1,445	1,643	1,525	1,510	1,619	1,392	1,428	1,464	1,620	1,666
8	1,425	1,629	1,512	1,499	1,614	1,381	1,416	1,453	1,617	1,659
9	1,409	1,618	1,501	1,491	1,611	1,372	1,405	1,444	1,614	1,652
10	1,395	1,609	1,492	1,483	1,607	1,364	1,396	1,435	1,611	1,646

Conclusion

The use of Rorschach ink-blots in the neurological as well as psychological and psychiatric sciences is a long debated proposition, a question of relating their 'meaning' as provided by patients, and clinical analysis thereof as an actual and accredited projective test.

Starting in 2009 we began a research project intended to investigate the actual structure of such ink-blots and we discovered they demonstrate a fractal structure.

By using the 2D-DFA procedure we pursued this notion and, in fact, in the present study we have obtained two results.

The first result is that we have demonstrated that the Rorschach ink-blots actually are multifractals. We have shown in detail that they may be described with accuracy using the standard multifractal variables and on this basis we may conclude that they are all strongly differentiated and may be ascribed a high level of complexity. The $\Delta\alpha$ and the Δf values discriminate in an excellent manner the ten figures also confirming the relative complexity of some figures in respect to the others.

As discussed in detail in the introduction, it is the high complexity of such figures which we hypothesize induce in the subject a large intrinsic and irreducible indetermination, representative of an inner conflict with an elevated level of emotional and cognitive dynamics in the subject. We consider that emotional responses and connected cognitive engagement are associated with the previous experiences of a subject, particularly relating to childhood and his/her early parental figures. Therefore we imply the particular relevance of these figures within a basic test.

The second step in the present investigation was the monitoring the GSR signal of the subjects while looking at each ink blot for two minutes, without asking them to express comments or interpretations but instead, to remain in silence while simply observing the ink blots. Obviously, in this experimental condition, subjective perceptive processes and hence, subject-specific individual mental/affective structure and individual cognitive responses were stimulated through each two minute observation spent in focused attendance of each ink blot.

Our 1D-DFA analysis of the recorded GSR revealed that the subject responds to the visual impression of the ink blots in neurological, psychological and mental dynamics that manifest a direct multifractal structure.

We estimated the $h(q)$ values for GSR Tone and Phase in each subject and, for all the ten Rorschach ink-blots. We obtained that such $h(q)$ values are discriminated for each ink blot. We reported also the value of the Hurst $h(2)$ exponent.

Our conclusion is that the high complexity of the Rorschach ink-blots activate corresponding highly complex neurological and mental dynamics that are estimated in the present paper by the variable $h(q)$.

From the viewpoint of basic and computational neuroscience, both of these results may be useful within the applied discipline of the Science of Complexity, as it augments and fleshes out our basic human neurological foundations and psychology.

The ability to estimate the index $h(q)$ in each subject relating the GSR in turn to each ink-blot, opens a new perspective of interest also for clinical applications, since for the first time we may apply a functional variable that connects Rorschach ink-blots with the GSR recording: defining the psychophysiological, neurological and mental condition of the subject, in an estimation of high accuracy. Therefore, the method may well permit quantitative evaluation of normal as well as pathological cases. Using basic science such as the theory of complexity in fundamental support, we believe this approach will be very helpful once clinically applied in order to to classify and study cases of pathological interest.

Appendix A

Step 1

Consider a gray image as a self-similar surface and represent it by an MxN matrix $X=X(i.j),\ i=1,2,....,M;\ j=1,2,....,N$. Partition the surface into a M_SxN_S non- overlapping square sub-surface of equal length$s$ where $M_S=M/s$ *and* $N_S=N/s$. Each sub-surface is denoted by $X_{m,n}=X_{m,n}(i,j)$ where $X_{m,n}(i,j)=X(r+i,t+j)$for $1 \leq i\ ,j \leq s$ and $r=(m-1)\ s$, $t=(n-1)\ s$.

Note that M and N are not necessarily multiples of the length s . Therefore, the sub-surfaces in the upper-right and the bottom may not be taken into consideration. We can then repeat the partitioning procedure starting from the other three corners.

Step 2

For each sub-domain $G_{m,n}$, find the cumulative sum as given in detail in (Wang et al., 2015)

Step 3

For each surface $G_{m,n}$ we can obtain a local trend $\overline{G}_{m,n}$ by fitting it with a pre-chosen bivariate polynomial function

$\overline{G}_{m,n}(i,j) = ai + bj + c$

And we can determine the residual matrix

$y_{m,n}(i,j) = G_{m,n}(i,j) - \overline{G}_{m,n}(i,j)$

Step 4

We calculate the detrended Fluctuation Function $F(m,n,s)$ of the segment $X_{m,n}$

$$F^2(m,n,s) = \frac{1}{s^2} \sum_{i=1}^{s} \sum_{j=1}^{s} y_{m,n}(i,j)^2$$

and the q-order Fluctuation Function $F_q(s)$

$$F_q(s) = \left[\frac{1}{M_s N_s} \sum_{m=1}^{M_s} \sum_{n=1}^{N_s} [F(m,n,s)]^q \right]^{1/q} \quad \text{for } q \neq 0$$

and

$$F_q(s) = \exp\left[\frac{1}{M_s N_s} \sum_{m=1}^{M_s} \sum_{n=1}^{N_s} \ln(F(m,n,s)) \right] \quad \text{for } q = 0$$

Step 5

Vary the value of s ranging from 6 to min (M,N)/4. In the cases in which the surfaces $G_{m,n}$ are long range power law correlated, $F_q(s)$ behaves, for large values of s, as a power law,

$F_q(s) \propto s^{h(q)}$

h(2) is the Hurst index h of the surface, and h(q) the generalized Hurst index of the surface. h(2) can be related to the fractal dimension D_f of the two-dimensional surface by means of the relationship $h = 3-D_f$.

The classical multifractal scaling exponents $\tau(q)$ are obtained by

$$\tau(q)=qh(q)-D$$

where D denotes the fractal dimension of the geometric support of the multifractal measure, for the two-dimensional measure, $D = 2$. It is also easy to obtain the generalized multifractal dimensions

$$D_q=\frac{\tau(q)}{q-1}=\frac{qh(q)-2}{q-1}$$

for $q \neq 1$ (for details see Wang et al., 2015)

In addition we have to characterize the multifractal surface via the singularity strength or Hölder exponent α and singularity spectrum $f(\alpha)$. They are related to $\tau(q)$ via a Legendre transform

$$\alpha(q)=\frac{d\tau(q)}{dq}$$

and

$$f(\alpha)=q\alpha(q)-\tau(q)$$

In addition we use the following indices of characterization:

a) $I = |h(q)-h(-q)|$(strength of multifractality) expressing the degree of multifractality.

b) $\Delta\alpha = \alpha_{max} - \alpha_{min}$

c) $\Delta f = f(\alpha_{max})-f(\alpha_{min})$ for all q respectively.

The larger the value of $\Delta\alpha$ is, the smaller the even distribution of probability measure is, and the more roughness image texture surface will usually be expected. The latter is Hausdorff dimension of the measured object, which is used to measure the degree of irregularity and complexity.

References

Abbound S., Berenfeld O., Sadeh D. (1991). Simulation of high-resolution QRS complex using ventricular model with a fractal conduction system. Effects of ischemia on high-requency QRS potentials. *Circ. Res.*, **68**: 1751–1760.

Atupelage C., Nagahashi H., Yamaguchi M., Sakamoto M., Hashiguchi A. (2012). Multifractal feature descriptor for histopathology. *Anal. Cell. Pathol.* (Amst.) **35**: 123–126.

Bassingthwaighte J., Van Beek J., King R. (1990). Fractal branchings: the basis of myocardial flow heterogeneities? *Ann. N. Y. Acad. Sci.,* **591**: 392–401.

Block A., von Bloh W., and Schnellhuber H.J. (1990). Efficient box-counting determination of generalised fractal dimensions. *Physical Review A,* **42**, 1869-1874.

Bullmore Ed., Barnes A., Bassett D.S., Fornito A., Kitzbichler M., Meunier D., Suckling J. (2009). Generic aspects of complexity in brain imaging data and other biological systems**.** *NeuroImage***, 47** (3),1125–1134.

Chhabra A. and Jensen R.V. (1989). Direct determination of the f(a) singularity spectrum. *The American Physical Society*, **62**, 1327-1330.

Conte E. (2015c). What Path Monitor: A Brief Note on Quantum Cognition and Quantum Interference, the Role of the Knowledge Factor. *Psychology*, **6**: 291-296.

Conte E., Farronato G., Farronato D., Maggipinto C., Maggipinto G., Zbilut J.P.(2008b). Recurrence quantification Analysis, Variability Analysis and Fractal Dimension Estimation in ^{99m}Tc-HDP Nuclear Scintigraphy of Maxillary Bones in subjects with Osteoporosis. *Chaos and Coomplexity Letters*, **3** (2), 111-142.

Conte E., Licata I., Alelù Paz R. (2015a) Formulation of the Quantum Neurological Model of Perception -Cognition and Awareness in Ambiguous Figures and in the Particular Case of the Dalmatian dog. *Journal of Behavioral and Brain Science* **5**(12): 533-549.

Conte E., Mastrolonardo M., Zbilut J.P. (2006). A fractal Analysis of Skin Pigmented lesions using the novel feature of the variogram. *Chaos, Solitons and Fractals*, **28**, 1119-1135.

Conte E., Norman R. (2016c). Some Brief Comments on Recent Studies on Quantum Cognition *NeuroQuantology* **14** (3):514-523.

Conte E., Norman R., Mordeniz C., Conte S., Marvulli R., Ianieri G., Megna M., Wang F. (2016b) A Multifractal Analysis of Neuron Locations and Calcium Activity in C-Elegans. *Chaos and Complexity Letters,* **10** (1): 1-23.

Conte E., Pieralice M., Tamburiello G., Battista E. (1991). Preliminary Results in mathematical, *Cybernetics*, IV/70: 2-14.

Conte E., Pierri G., Federici A., Mendoloccio L, Zbilut J.P. (2009). The Fractal variogram Analysis as General Tool to Measure Normal and Altered Metabolism States and the

Genetic Instability: an application to the case of the cutaneous malignat melanoma. *Chaos and Complexity Letters*, **3** (3), 273-280.

Conte E., Rui Freire L. (2015b) First Time Demonstration of the Quantum Interference Effect during Integration of Cognition and Emotion in Children. *World Journal of Neuroscience*, **5**: 91-98.

Conte E., Todarello O., Mendolicchio L., Corfiati L., Macina A., Pieri G., Zbilut J.P. (2008a). A Psycho-physical Model of of Rorschach's ink-blots using fractal analysis with estimation of the generalized fractal dimension and fractal variance function. *Chaos and Complexity Letters*, **4** (1), 77-84.

Conte E., Todarello O., Mendolicchio L., Corfiati L., Macina A., Federici A., Pierri G., Zbilut J.P. (2012a). A Psycho-physical Model of of Rorschach's ink-blots using fractal analysis with estimation of the generalized fractal dimension and fractal variance function. *On the Borders of Complexity*. Nova Science Publishers Inc.

Conte E., Wang F., Norman R., Mordeniz C., Conte S., Licata I. (2016a) A Multifractal Analysis of Connectomics, Neuron Locations, And Calcium Activity In C-Elegans. *Chaos and Complexity Letters,* **10** (2): 111-127.

Conte E. (2011a). On the Logical Origins of Quantum Mechanics Demonstrated by Using Clifford Algebra: A Proof That Quantum Interference Arises in a Clifford Algebraic Formulation of Quantum Mechanics. *Electronic Journal of Theoretical Physics*, **8**: 109-126.

Conte E. (2011b). On the Logical Origins of Quantum Mechanics Demonstrated by Using Clifford Algebra. *Neuroquantology*, **9**: 231-242.

Conte E. (2011c). *Advances in Application of Quantum Mechanics in Neuroscience and Psychology: A Clifford Algebraic Approach.* New York: Nova Science Publishers.

Conte E. (2013a). Are Information, Cognition, and the Principle of Existence Intrinsically Structured in the Quantum Model of Reality? *Advanced Studies in Theoretical Physics*, **7**: 797-818.

Conte E. (2013b). A Clifford Algebraic Analysis Gives Mathematical Explanation of Quantization of Quantum Theory and Delineates a Model of Quantum Reality in Which Information, Primitive Cognition Entities and a Principle of Existence Are Intrinsically Represented Ab Initio. *World Journal of Neuroscience*, **3**: 157-170.

Conte E., Khrennikov, A.Y., Todarello, O., De Robertis, R., Federici, A., & Zbilut, J.P. (2011d). On the Possibility That We Think in a Quantum Mechanical Manner: An Experimental Verification of Existing Quantum Interference Effects in Cognitive Anomaly of Conjunction Fallacy. *Chaos and Complexity Letters*, **4:** 123-136.

Conte E., Khrennikov, A.Y., Todarello, O., Federici, A., Mendolicchio, L., & Zbilut, J.P. (2009b). Mental States Follow Quantum Mechanics during Perception and Cognition of Ambiguous Figures. *Journal of Open Systems and Information Dynamics*, **16**: 1-17.

Conte E., Khrennikov, A.Y., Todarello, O., Federici, A. and Zbilut, J.P. (2009c) On the existence of quantum wave function and quantum interference effects in mental states: An experimental confirmation during perception and cognition in humans. *Neuroquantology*, **7**: 204-212.

Conte E., Khrennikov, A.Y., Todarello, O., Federici, A. and Zbilut, J.P. (2008c) A preliminary experimental verification on the possibility of bell inequality violation in mental states. *Neuroquantology*, **6**: 214-221.

Conte E., Santacroce, N., Laterza, V., Conte, S., Federici, A., & Todarello, O. (2012b). The Brain Knows More Than It Admits: A Quantum Model and Its Experimental Confirmation. *Electronic Journal of Theoretical Physics*, **9**: 72-110.

Conte E., Todarello, O., Federici, A., Vitiello, F., Lopane, M., Khrennikov, A.Y., & Zbilut, J.P. (2007). Found Experimental Evidence of Quantum Like Behavior of Cognitive Entities. An Abstract Quantum Mechanical Formalism to Describe Cognitive Entities and Its Dynamics. *Chaos, Solitons and Fractals*, **3**: 1076-1088.

Conte E., Todarello, O., Laterza, V., Khrennikov, A.Y., Mendolicchio, L. and Federici, A. (2010) A preliminary experimental verification of violation of bell inequality in a quantum model of Jung theory of personality formulated with Clifford algebra. *Journal of Consciousness Exploration & Research*, **1**: 831-849.

Cornforth D. and Jelinek H.F. (2005). Automated classification of dementia subtypes from post-mortem cortex images. In: AI2005: *Advances in Artificial Intelligence*, pp. 1285-1288, Sydney, Australia.

Delignières D., Fortes M.,Ninot G., (2004), The fractal dynamics of Self-Esteem and physical self, *Non Linear dynamics in Psychology and LifeSciences,* 8,479-510.

Di Leva A., Esteban F.J., Grizzi F., Klonowski W., Landrove M.M. (2015). Fractals in the Neurosciences, Part II: Clinical Applications and Future Perspectives. *The Neuroscientist*, **21**(1), 30-43.

Fernández E. and Jelinek H.F. (2001). Use of fractal theory in neuroscience: methods, advantages, and potential problems. *Methods*, **24**, 309-321.

Fernández E., Bolea J.A., Ortega G., and Louis E. (1999). Are neurons multifractals? *Journal of Neuroscience Methods*, **89**, 151-157.

Fu-Zen Shaw, Ruei-Feng Chen, Hen-Wai Tsao, Chen-Tung Yen (1999). Algorithmic complexity as an index of cortical function in awake and pentobarbital-anesthetized rats**.** *Journal of Neuroscience Methods*, **93** (2), 101–110.

Gisiger T. (2001). Scale invariance in biology: coincidence or footprint of a universal mechanism? *Biological Reviews of the Cambridge Philosophical Society,* **76** (2): 161-209.

Goldberger A.L. (1996). Non-linear dynamics for clinicians: chaos theory, fractals, and complexity at the bedside. *Lancet,* **347**: 1312–1314.

Goldberger A.L., Amaral L.A., Hausdorff J.M., Ivanov P.C., Peng C.K., Stanley H.E. (2002). Fractal dynamics in physiology: alterations with disease and aging. *Proc. Natl. Acad. Sci. U.S.A.*, **99**: 2466–2472.

Gu G.F., Zhou W.X. (2006). Detrended fluctuation analysis for fractals and multifractals in higher dimensions. *Phys Rev E.*, **74**: 061104.

Hausdorff J.M. (2007). Gait dynamics, fractals and falls: finding meaning in the stride-to-stride fluctuations of human walking. *Hum. Mov. Sci.*, **26**: 555–589.

Iannaccone P.M. and Khokha M. (1995). Fractal geometry in biological systems: an analytical approach, CRC Press, New York.

Ivanov P.C., Amaral L.A., Goldberger A.L., Havlin S., Rosenblum M.G., Struzik Z.R., and Stanley H.E. (1999). Multifractality in human heartbeat dynamics. *Nature,* **399**: 461–465.

Jelinek H., Karperien A., Bossomaier T., and Buchan A. (2004b). Differentiating grades of microglia activation with fractal analysis. *Proceedings of the 7th AsiaPacific Conference on Complex Systems, pp. 605-612, Cairns, Australia.*

Jelinek H.F. and Fernández E. (1998). Neurons and fractals: how reliable and useful are calculations of fractal dimension? *Journal of Neuroscience Methods*, **81**, 9-18.

Jelinek H.F., Cornforth D.J., Roberts T., Landini G., Bourke P., and Bossomaier T. (2004a). Image processing of finite size rat retinal ganglion cells using multifractal and local connected fractal analysis. In: *Complex Systems Conference*, Cairns, Australia.

Jelinek H.F., Steinke A.B. (1996). Determination of the fractal dimension of cat retinal ganglion cells using a new method on the World Wide Web. *Proc Austr Neurosci Soc*, **7**:139-146.

Jestczemski F. and Sernetz M. (1996). Multifractal approach to inhomogeneous fractals. *Physica A*, **223**, 275-282.

Jian Li, Qian Du, Caixin Sun (2009). An improved box-counting method for image fractal dimension estimation. *Pattern Recognition*, **42** (11), 2460–2469.

Kantelhardt J.W., Zschiegner S.A., Koscielny-Bunde E. et al. (2002). Multifractaldetrended fluctuation analysis of nonstationary time series. *Phys. A Stat. Mech.Appl.*, **316**, 87–114.

Karperien A., Ahammer E., Jelinek H.F. (2013). Quantitating the subtleties of microglial morphology with fractal analysis. *Front Cell Neurosci.*, **7**: 3.

Krenz G., Linehan J., Dawson C. (1992). A fractal continuum model of the pulmonary arterial tree. *J. Appl. Physiol.*, **72**: 2225–2237.

Losa G. (2014). On the fractal design of human brain and Nervous Tissue. *Applied mathematics*, **5**, 1725-1732.

Losa G. (2015). The living realm depicted by the fractal geometry. *Fractal Geometry and Nonlinear Anal in Med and Biol*, **1**(1), 11-15.

Mandelbrot B.B. (1983). The fractal geometry of nature. Freeman and Co, New York.

Mattei T.A. (2014). Unveiling complexity: non-linear and fractal analysis in neuroscience and cognitive psychology. *Frontiers in Computational Neureoscience*, **8**, article 17.

Mitina O.V., Abraham F.D. The use of fractals for the study of the psychology of perception, http://www.blueberry-brain.org/dynamics/mitina-fractal-perception.htm.

Nazahah M., Trevor S. Ahearn, Gordon D. Waiter, Alison D. Murray, Lawrence J. Whalley, Roger T. Staff (2012). Brain structural complexity and life course cognitive change. *NeuroImage,* **61**, 694–701.

Nonnenmacher T.F., Baumann G., Barth A., and Losa G.A. (1994). Digital image analysis of self-similar cell profiles. *International Journal of Bio-Medical Computing*, **37**, 131-138.

Norman R., Mendolicchio L., Mordeniz C., Pierangeli E., Pannarale P., Orsucci F., Conte E.(2016a). Galvanic skin response and its neurological correlates, *Journal of Consciousness Exploration and Research,* 7,1,553-572; Mind Science ViXra .org 1606:0140;1606:0095.

Norman R., Mordeniz C, Mendolicchio L., Pierangeli E., Pannarale P., Orsucci F., Conte E. (2016b). A basic biofeedback primer, Mind Science, ViXra.org, 1606:0140.

Norman R.L. (2016). Super-ego & the Neuroscience of Empathy: From Unconscious Wish to Manifest Behavior - A New Human Model. *Journal of Consciousness Exploration and Research*, Vol. 7 (9) 677-697 ISSN: 2153-8212.

Orsucci F., Petrosino R., Paoloni G., Canestri L., Conte E., Reda M.A.,(2013) Fulcheri M., Prosody and synchronization in cognitive neuroscience, *EPJ Non Linear Biomedical Physics*, **1** (6): 1-11.

Parkinson I., Fazzalari N. (1994). Cancellous bone structure analysis using image analysis. *Australas. Phys. Eng. Sci. Med.*, **470**: 64–70.

Peng C.K., Mietus J.E., Liu Y., Lee C., Hausdorff J.M., Stanley H.E.,., Goldberger A.L., Lipsitz L.A. (2002). Quantifying fractal dynamics of human respiration: age and gender effects. *Ann. Biomed. Eng.*, **30**: 683–692.

Peng C-K, Buldyrev S.V., Havlin S, Simons M, Stanley H.E., Goldberger A.L. (1994). Mosaic organization of DNA nucleotides. *Phys Rev E*, **49**, 1685-1689.

Ristanović D., Stefanović B.D., Milošević N.T., Grgurević M., and Štulić V. (2002). Fractal and nonfractal analysis of cell images: comparison and application to neuronal dendritic arborization. *Biological Cybernetics*, **87**, 278-288.

Ristanovich D., Losa G.A. (2013). A contribution to definitions of some fractal concepts. *The Fractal Laboratory Journal*, **2**, 2.

Schroeder M. (1991). *Fractals, chaos, power laws.* W.H Freeman and Co, New York.

Schroeder M., (1991), Fractals, Chaos, Power Laws: Minutes from an Infinite Paradise, W.H. Freeman & Co, 1991 New York.

Smith Jr T.G., Lange G.D., and Marcks W.B. (1996). Fractal methods and results in cellular morphology - dimensions, lacunarity and multifractals. *Journal of Neuroscience Methods*, **69**, 123-136.

Smith Jr T.G., Lange G.D., and Marcks W.B. (1996). Fractal methods and results in cellular morphology - dimensions, lacunarity and multifractals. *Journal of Neuroscience Methods*, **69**, 123-136.

Stanley H.E., Amaral L.A.N., Goldberger A.L., Havlin S., Ivanov P.Ch., and Peng C.K. (1999). Statistical physics and physiology: Monofractal and multifractal approaches. *Physics A*, **270**, 309-324.

Vicsek T. (1992). Fractal Growth Phenomena, World Scientific, Singapore.

Voss R.M. and Wyatt J.C.Y. (1991). Multifractals and local connected fractal dimension: Classification of early chinese landscape paintings. In: *Applications of fractals and chaos* (A. J. Crilly, R.A. Earnshaw and H. Jones. (Eds)), Springer-Verlag, Berlin.

Wang C., Allegaert K., Peeters M.Y.M., Tibboel D., Danhof M., and Knibbe C.A.J. (2014). The allometric exponent for scaling clearance varies with age: a study on seven propofol datasets ranging from preterm neonates to adults. *British Journal of Clinical Pharmacology,* **77** (1) 149–159.

Wang F., Deng wen Liao, Jin wen Lii, Giung -Ping Jao (2015). Two-dimensional Multifractal detrended fluctuation analysis of plant identification. *Plant Methods*, **11** (12), 1-12.

Wang F., Li J.W., Shi W., Liao G.P. (2014a). Leaf image segmentation method based on multifractal detrended fluctuation analysis. *J Appl Phys.*, **114** (21):214905.

Wang F., Li Z.S., Liao G.P. (2014b). Multifractal detrended fluctuation analysis for image texture feature representation. *International Journal of Pattern Recognition and Articial Intelligence,* **28** (3),1-20.

Wang G., Huang H., Xie H., Wang Z., Hu X. (2007). Multifractal analysis of ventricular fibrillation and ventricular tachycardia. *Med. Eng. Phys.*, **29**: 375–379.

Weibel E.R. (1991). Fractal geometry: a design principle for living organisms. *Am. J. Physiol.,* **261**: 361–369.

Zheng Y., Gao J.B., Sanchez J.C., Principe J.C., Okun M S. (2005). Multiplica-tive multifractal modeling and discrimination of human neuronal activity. *Phys. Lett. A,* **344**: 253–264.

Index

R

S

T

V